SCIENCE FOR DECISIONMAKING

Coastal and Marine Geology at the U.S. Geological Survey

Ocean Studies Board

Commission on Geosciences, Environment, and Resources

National Research Council

NATIONAL ACADEMY PRESS
Washington, D.C.

NATIONAL ACADEMY PRESS • 2101 Constitution Avenue, NW • Washington, DC 20418

NOTICE: The project that is the subject of this report was approved by the Governing Board of the National Research Council, whose members are drawn from the councils of the National Academy of Sciences, the National Academy of Engineering, and the Institute of Medicine. The members of the committee responsible for the report were chosen for their special competences and with regard for appropriate balance.

This report and the committee were supported by a grant from the United States Geological Survey. The views expressed herein are those of the authors and do not necessarily reflect the views of the sponsor.

International Standard Book Number 0-309-06584-4

Additional copies are available from the National Academy Press, 2101 Constitution Ave., NW, Washington, DC 20418 (1-800-624-6242; http://www.nap.edu).

Printed in the United States of America

COMMITTEE TO REVIEW THE USGS COASTAL AND MARINE GEOLOGY PROGRAM

JOAN OLTMAN-SHAY, *Chair*, Northwest Research Associates, Inc., Bellevue, Washington
JAMES COLEMAN, Louisiana State University, Baton Rouge
ARTHUR GREEN, Exxon Exploration Company, Houston, Texas
SUSAN HUMPHRIS, Woods Hole Oceanographic Institution, Massachusetts
CURT MASON, National Oceanic and Atmospheric Administration, Silver Spring, Maryland
NEIL OPDYKE, University of Florida, Gainesville
NANCY RABALAIS, Louisiana Universities Marine Consortium, Chauvin
NOEL TYLER, Texas Bureau of Economic Geology, Austin

Staff

DAN WALKER, Study Director
SHARI MAGUIRE, Senior Project Assistant
JODI BACHIM, Project Assistant

OCEAN STUDIES BOARD

COMMISSION ON GEOSCIENCES, ENVIRONMENT, AND RESOURCES

The National Academy of Sciences is a private, nonprofit, self-perpetuating society of distinguished scholars engaged in scientific and engineering research, dedicated to the furtherance of science and technology and to their use for the general welfare. Upon the authority of the charter granted to it by the Congress in 1863, the Academy has a mandate that requires it to advise the federal government on scientific and technical matters. Dr. Bruce Alberts is president of the National Academy of Sciences.

The National Academy of Engineering was established in 1964, under the charter of the National Academy of Sciences, as a parallel organization of outstanding engineers. It is autonomous in its administration and in the selection of its members, sharing with the National Academy of Sciences the responsibility of advising the federal government. The National Academy of Engineering also sponsors engineering programs aimed at meeting national needs, encourages education and research, and recognizes the superior achievements of engineers. Dr. William A. Wulf is president of the National Academy of Engineering.

The Institute of Medicine was established in 1970 by the National Academy of Sciences to secure the services of eminent members of appropriate professions in the examination of policy matters pertaining to the health of the public. The Institute acts under the responsibility given to the National Academy of Sciences by its congressional charter to be an adviser to the federal government and, upon its own initiative, to identify issues of medical care, research, and education. Dr. Kenneth I. Shine is president of the Institute of Medicine.

The National Research Council was organized by the National Academy of Sciences in 1916 to associate the broad community of science and technology with the Academy's purposes of furthering knowledge and advising the federal government. Functioning in accordance with general policies determined by the Academy, the Council has become the principal operating agency of both the National Academy of Sciences and the National Academy of Engineering in providing services to the government, the public, and the scientific and engineering communities. The Council is administered jointly by both Academies and the Institute of Medicine. Dr. Bruce Alberts and Dr. William A. Wulf are chairman and vice-chairman, respectively, of the National Research Council.

Preface

The nation's coastal and marine areas are some of our greatest environmental assets. However, these areas are under pressure from increasing coastal population and resource demands. In order to manage these areas wisely, decision-makers are demanding more and more information about natural marine systems and the impacts of human activities on them. Thus, when the U.S. Geological Survey approached the Ocean Studies Board (OSB) with the request to review the Coastal and Marine Geology Program, the value of such a review needed no explanation.

The OSB established the Committee to Review the USGS Coastal and Marine Geology Program and charged us with recommending areas of focus for the survey's future activities in the coastal and marine regions, strategies for hiring and retaining high-quality staff, and measures for most effectively directing fiscal and human resources toward the unique challenges that exist in these important regions. This report reflects the conclusions and recommendations of the committee, drawing on extensive discussions with USGS staff; input from potential users, clients, and collaborators of the Coastal and Marine Geology Program (CMGP); and the committee's own expertise and experience.

Throughout this report the reader will find short descriptions of relevant studies conducted by the CMGP. These examples are a small subset of the large number of studies conducted by CMGP in recent years. Many other examples can be found at the CMGP homepage *http://marine.usgs.gov/*.

The Committee to Review the USGS Coastal and Marine Geology Program is very grateful to the many individuals who played a significant role in the completion of this study. The committee met four times and extends its gratitude to the following individuals who appeared before the full committee or otherwise provided background information and discussed pertinent issues: Peter Barnes,

Brad Barr, Peter Betzer, Steven Bohlen, Mike Bothner, Kenneth Brink, Brad Butman, Michael Carr, William Dillon, Carlton Dufrechou, Terry Edgar, Michael Fields, Laura Fredricks, Robert Gagosian, James Gardner, Leon Gove, Gary Griggs, Chip Groat, Bob Halley, Monty Hampton, Janet Hashimoto, Deborah Hutchinson, George Kaminsky, Robert Kayen, Jack Kindinger, Randall Koski, Patrick Leahy, Homa Lee, Ralph Lewis, Jeff List, Patrick Muffler, Bruce Richmond, Dave Russ, Abby Sallenger, Rex Sanders, William Schwab, Gene Shinn, Marilyn ten-Brink, Uri ten-Brink, Rob Wertz, Jeff Williams, and Richard Williams.

This report was reviewed in draft form by individuals chosen for their diverse perspectives and technical expertise, in accordance with procedures approved by the NRC's Report Review Committee. The purpose of this independent review is to provide candid and critical comments that assist the institution in making the published report as sound as possible and to ensure that the report meets institutional standards for objectivity, evidence, and responsiveness to the study charge. The review comments and draft manuscript remain confidential to protect the integrity of the deliberative process. We wish to thank the following individuals for their participation in the review of this report: Jeff Benoit (National Oceanic and Atmospheric Administration), Steven Boss (University of Arkansas), James Funk (Shell Oil), Eldon Hout (Oregon State Coastal and Ocean Management Program), Ray Krone (University of California at Davis), Rutherford Platt (University of Massachusetts at Amherst), Andrew Solow (Woods Hole Oceanographic Institution), Debra Stakes (Monterey Bay Aquarium Research Institute), and Walter Schmidt (Florida Geological Survey). While the individuals listed above have provided constructive comments and suggestions, it must be emphasized that responsibility for the final content of this report rests entirely with the authoring committee and the institution.

Finally, the committee wishes to gratefully acknowledge the efforts of the Ocean Studies Board staff who helped produce this report, particularly, the study director, Dan Walker, and the project assistants, Shari Maguire and Jodi Bachim. Without their guidance and help, this project could never have been completed.

JOAN OLTMAN-SHAY
Chair, Committee to Review the USGS
Coastal and Marine Geology Program

Contents

Executive Summary

The coastal and marine areas of the United States represent some of the most diverse and resource-rich in the nation. The abundant resources in these areas and their aesthetic beauty make them attractive areas to work, live, and play. However, to a very large extent these are also extremely fragile ecosystems; thus, the very attributes that have made them an ever-increasing focus of life in the United States make these regions and the resources they contain extremely vulnerable to mismanagement.

The ability to manage these areas wisely and to position society to reap the maximum sustainable benefit of their resources will lie in a scientific understanding of the processes that control the distribution and functioning of this enormous wealth. As has been recognized widely in the ocean science community for several years, the complex nature of many of the coastal and marine policy issues facing decisionmakers at the federal, state, and local levels transcends the disciplinary boundaries that have characterized scientific research over the last 200 years. Over the last decade, the terms *interdisciplinary research* and *systems science* have become commonplace, yet effective execution of the concepts remains difficult. Furthermore, with more and more attention placed on coastal and marine areas, the number of federal agencies involved and the overlapping roles of these entities, as well as state agencies and academic institutions, have created an extremely complex state of affairs.

As part of an ongoing effort by the Geologic Division of the United States Geological Survey (USGS) to receive input from the broader scientific community, the division requested that the National Research Council (NRC) conduct reviews of a number of its ongoing programs, including the Coastal and Marine

Geology Program (CMGP). In response to this request, the NRC formed the Committee to Review the USGS Coastal and Marine Geology Program. The committee was asked to review the history and status of the CMGP, particularly its most recent national plan and recent workshop reports in the context of the USGS and the new Geologic Division's science strategy, and provide advice on:

- the general areas of future program emphasis (e.g., research, national assessments, monitoring, characterization) and cooperation with local, state, and national decisionmakers and with government and academic scientists;
- the specific scientific and technical challenges (including components from the national plan such as coastal erosion, earthquake hazards, pollution studies, biologic habitats, distribution and significance of gas hydrates), as well as the challenge to maintain a strong and dedicated research staff;
- balancing between issue-driven and knowledge-driven research and balancing between regional and national efforts;
- the ideal mix of science staff as to discipline and status (permanent versus term) to meet needs and ensure long-term health of the Coastal and Marine Geology Program; and
- the ideal ratio of core-funded research versus reimbursable research paid by clients.

This report, the outcome of the committee's review, attempts both to provide an understanding of the importance of the geologic sciences in understanding the coastal and marine areas of the United States and to provide advice to the USGS about how to better focus its efforts in this regard.

THE VALUE OF UNDERSTANDING GEOLOGIC PROCESSES AND THE ROLE OF THE U.S. GEOLOGICAL SURVEY

The coastal and marine areas of the United States not only include the landward portion of the region popularly thought of as the coast but state waters (which commonly extend offshore 3 nautical miles) and what is referred to as the U.S. Exclusive Economic Zone (EEZ). The EEZ, which extends seaward 200 nautical miles from the coastline, covers an area of 3 million square nautical miles (an area 30 percent larger than the land area of the entire United States). These coastal and marine regions owe much of their unique character to the geologic processes that formed the continent of North America and various islands of the Pacific and Atlantic oceans.

These geologic processes, in concert with atmospheric and oceanic processes, control the elevation of coastal areas, the bathymetry of the coastal seas and oceans, and the location of many of the commonly recognized features of these unique areas. These same processes, which moderate and interact with ecosystems, control the distribution of mineral and water resources, the patterns of shoreline change, the extent and nature of wildlife habitat, and the living marine resources that support a substantial segment of the U.S. economy. Simply put, the coasts exist because of the geologic forces that formed the continents, islands, and oceans that cover Earth.

Thus, wise stewardship and development of many coastal and marine natural resources are linked to sound scientific understanding. Science-based policy decisions facing federal, state, and local policymakers can be expected to depend on an understanding of the processes that have traditionally been the focus of research by the USGS and its CMGP.

Although several federal agencies conduct physical science and engineering programs and studies, the CMGP occupies a unique niche by providing the capability to conduct research and assessments of the geologic processes impacting the nation's coasts. The U.S. Army Corps of Engineers focuses on developing engineering solutions to very site-specific coastal problems (e.g., tidal inlet improvement projects and beach nourishment projects). The National Oceanic and Atmospheric Administration's needs for geologic information to address its mission requirements for management of fisheries, sanctuaries, and other coastal resources are not met in the agency, although the Sea Grant program does support small geologic research studies conducted by state institutions. The U.S. Federal Emergency Management Agency (FEMA) and the U.S. Environmental Protection Agency (EPA) rely heavily on the academic community to provide whatever geologic research and knowledge base they require. However, the USGS alone has the ability to frame coastal geologic questions having both regional and national perspectives, while conducting studies that provide the geologic component for interdisciplinary approaches and useful information to decisionmakers. The distinctly different geologic characteristics of the coastal and marine realm of the United States, as well as the variations in ocean circulation and weather patterns, result in different geologic processes with diverse spatial and temporal scales that shape the coastlines and seafloor. The CMGP is uniquely qualified to address these issues given its capability both to conduct nearshore and offshore marine geologic studies and to integrate the results to produce a national assessment of the geologic structure of the coastal areas and adjacent EEZ.

FUTURE CHALLENGES

The committee identified the major scientific questions or *grand challenges* that should form the integrating principle common to all CMGP efforts to fulfill the need for geological information about the nation's coastal and marine areas over the next few decades. To respond adequately to these grand challenges the CMGP will need to consider changes in the existing CMGP structures and procedures.

The three grand challenges identified by the committee are intended to provide a long-term focus and are not site or issue specific. These grand challenges include: 1) establish the geologic framework of the U.S. coastal and marine regions; 2) develop a national knowledge bank on the geologic framework of these regions; and 3) develop a predictive capability based on an understanding of the geologic framework of these regions. These challenges are intended as an integrative principle that should be used to evaluate the relevance of a variety of projects over the next 10 to 15 years (or longer). The resulting investigative program will be varied, as the complexity of the nation's coastal and marine areas varies spatially, and the underlying need for information will vary temporally. **Successful execution of a national investigative program will require a *systems-science* approach (broad interdisciplinary and integrated studies) rather than single-discipline-based or geographically localized projects.** In addition, **addressing these challenges will require the CMGP to make greater use of expertise that may reside in other USGS units, federal or state agencies, or academic institutions.** Such expanded interactions should enable CMGP to better communicate the results of its efforts to its user community.

ROLE OF THE CMGP IN MEETING THE CHALLENGE

The committee believes that CMGP, by organizing activities at all three regional centers through an integrated plan to address the grand challenges discussed in Chapter 3, would be well positioned to meet the nation's need to address national, regional, and site-specific coastal and marine issues and problems. A well-crafted vision statement will define goals that, when coupled with a thoughtful strategic plan, are relevant to the actions of every CMGP staff member and to every action undertaken by the CMGP. **The committee therefore recommends that CMGP leadership initiate a program-wide strategic planning process to establish goals and objectives for integrated science efforts. As part of this strategic planning, a new mission statement should be developed that identifies the role of the CMGP and its responsibilities to the nation.** Such a statement should reflect the responsibilities of the CMGP:

I. to conduct research to advance our understanding of the dynamic processes, both natural and anthropogenic, which change the coastlines and seafloor along coastal margins;
II. to provide the geologic framework for policy decisions regarding the use and management of the marine environment but also to respond to the needs of other federal, state, and local agencies when coastal geologic data and assessments are required to address critical management and policy issues; and
III. to provide information critical to planning for the future environmental and economic health of the nation's coastal areas, including an understanding of the likely scenarios for change to the geologic framework of coastal environments, whether from long-term climate change or rapid changes from extreme events or human activities.

IMPLEMENTING CHANGE

Timely input and guidance from the CMGP's clients and collaborators will be crucial to the successful use of the CMGP's limited resources to serve the nation's need for scientific understanding of coastal and marine geologic processes. **Consequently, the committee recommends that the present Program Council be replaced with an Advisory Council charged with new responsibilities and constituted to reflect the need for broad input to the CMGP.**

It is generally recognized that the present USGS staff is talented and uniquely positioned to identify major relevant scientific challenges and design research strategies to address them. The ability to recruit and maintain a high-quality staff will depend on identifying ways to reward creative and resourceful personnel. A long-term commitment to a robust and focused research strategy should encourage staff to make a similar commitment to the program, reducing turnover while encouraging potential program staff to join the USGS effort. **The committee recommends that CMGP leadership, during its strategic planning effort, identify the disciplines that will be required by the CMGP to meet its long-term goals. Ensuring that these discipline areas are well represented during subsequent hiring efforts should be a priority. Furthermore, because these efforts should reflect long-term needs, care should be made to hire at a consistent and even rate.**

Realization of the long-term goals represented by the grand challenges, as well as the near-term objectives, will greatly depend on CMGP's ability to continue to develop collaborative relationships with other federal, state, and local agencies, as well as academia. **CMGP should make every effort to leverage the expertise found in the government (including other programs and divisions of the USGS), academia, and private industry to expand its ability to meet the needs of its diverse user community.**

1

Introduction

The coastal regions of the United States are economically vital areas, supporting diverse industries and large population centers. When considered in combination with the adjacent marine areas that comprise the U.S. Exclusive Economic Zone (EEZ), they represent one of the greatest environmental assets of the nation. For example, the EEZ extends seaward 200 nautical miles from our coastline and covers an area of 3 million square nautical miles (an area 30 percent larger than the land portion of the United States [Gardner et al., 1996]). The coastal and marine regions of the United States encompass vast and complex environments from terrestrial to the air-sea-land interface to the coastal ocean, the continental margin, and the deep ocean and are characterized by rich biological diversity and a wealth of mineral resources. As society has increasingly populated the coasts over the last 25 years, recreated to the beaches, dammed the rivers feeding the beaches and coasts, harvested fish, disposed of waste, and used these areas for transportation, the natural health of the coastal and marine environment has become a critical issue.

The beauty of the coastal ocean has drawn more and more people to inhabit and heavily use this delicately balanced area. The population in U.S. coastal counties currently exceeds 141 million (U.S. Bureau of the Census, 1998). These coastal counties account for only 17 percent of the U.S. landmass; thus, over half of the U.S. population lives in less than one-fifth of its total area, and this trend is expected to continue. For example, 17 of the 20 fastest-growing counties are located along the coast (NOAA, 1998). Nearly 14,000 new housing units are built in coastal counties every week (NOAA, 1998). Beaches have become one of the largest vacation destinations in America, with 180 million people visiting the coast every year (Cunningham and Walker, 1996). This increase in recreational usage, together with the impact of larger year-round populations, is stressing available resources and is making the safe and prudent management of these areas increasingly challenging. Coastal ecosystems face a variety of major environmental problems, including habitat modification, degraded water resources, toxic contamination, introduction of non-indigenous species, and shoreline erosion and vulnerability to storms and tsunamis.

Although recent improvements in hurricane, El Niño, and severe coastal storm forecasting have sharply reduced loss of life, the ongoing shift in U.S. population to the coastlines has resulted in an increase in risk to property and human life caused by storms and other geologic processes characteristic of coastal settings, such as earthquakes, landslides, and coastal erosion. Estimates of the present total value of insured property at risk range from $2 to $3.15 trillion (Lewis and Murdock, in press). In addition, millions of individuals are now at risk from rapid-onset events, such as tsunamis, for which present forecasting and early warning capabilities may be less effective. The rapidly increasing expenditures associated with relief and recovery from coastal disasters are of growing concern to both the federal government and the nation as a whole. These coastal disasters result when human behavior and natural processes combine to place homes, businesses, and the public infrastructure at risk.

In many coastal communities, declining groundwater levels have led to saltwater intrusion in previously pristine aquifers. Freshwater and saltwater flows are known to form a dynamic system on the continental margin. For example, freshwater springs have long been known off the southeast U.S. coast, and brine seeps have enabled unusual chemosynthetic biological communities to develop in deep water on the continental margin off Florida, California, and Alaska. Saltwater intrusion into freshwater aquifers in coastal areas illustrates that human activities can alter the flow unfavorably. Along the East Coast, onshore and offshore coastal aquifers form essentially contiguous regional units between Rhode Island and Florida. Little is known, however, about the details of the distribution, hydrology, and volume of freshwater in the coastal and offshore region or the extent of its connection with onshore aquifers. Understanding the controls that subsurface geology places on aquifer characteristics is critical to wise use of groundwater. How water (and other fluids, such as waste) flows through continental margin deposits, its interaction with the host sediments, and the relative role that changing sea level plays are often poorly known.

The effects of coastal problems ripple far beyond the coastal lands and beaches out on to the continental shelf and slope and even into the deep ocean. Poorly planned development in these areas can result in destruction of the salt marshes and wetlands that act as nurseries for many fish stocks. The damming of rivers can reduce the supply of sediment to beaches, accelerating coastal erosion. Pollutants can be transported either dissolved in the seawater or on particles that can carry these materials far offshore where they can affect the marine ecosystems that sustain our fisheries (Plate 1). These processes, coupled with over fishing and the use of fishing techniques that are non-species-specific or physically damage habitats are raising questions about the sustainability of the nation's fisheries. Key to the protection and wise use of these resources is an understanding of the role of habitat in population dynamics, which are often dependent on the geologic framework and geologic and ocean processes in local areas.

The deep marine realm, beyond the continental shelf and out to the edge of the EEZ (Fig. 1-1), is commonly regarded as a distant region, unrelated to human endeavor; hence, it is often dismissed in terms of its impact on human life. However, many ostensibly coastal issues have an offshore component. The diversity of the nation's coastal and marine environments is in part due to offshore geologic processes. For example, off Oregon and Washington, volcanic and earthquake activity is concentrated near the edge of the EEZ along a mid-ocean ridge system, which is the longest volcanic mountain chain on our planet (Box 1-1). In this area, hydrothermal vents, similar to the more familiar hot springs of Yellowstone National Park, discharge hot fluids onto the seafloor; form mineral deposits rich in iron, copper, and zinc; and are the site of exotic biological communities. Enzymes from the bacteria found at these sites are now being used extensively in the biotechnology industry and are benefiting humankind through biomedical research. In contrast, the Gulf of Mexico has been dominated

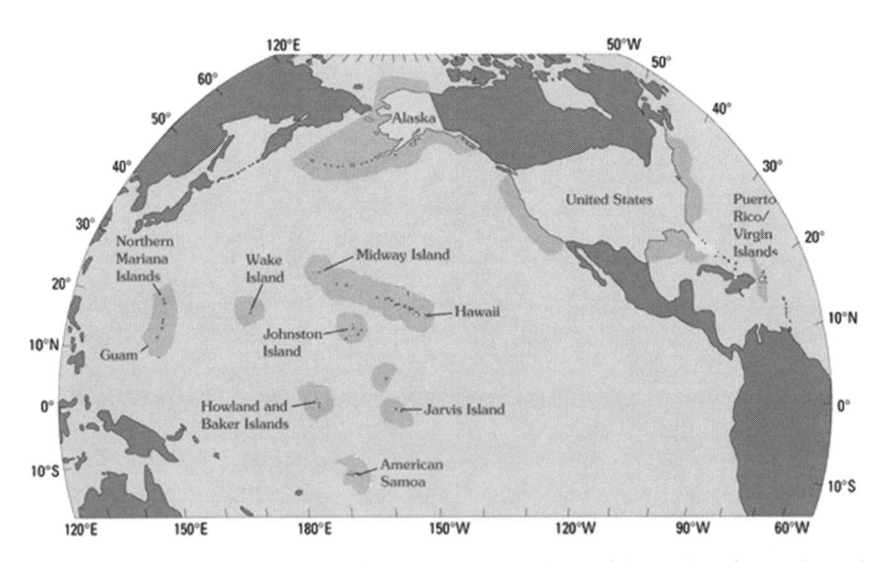

FIGURE 1-1 U.S. Department of the Interior holdings with coral reefs; total coral reef acreage about 625,000 acres. Shaded area is the Exclusive Economic Zone of the United States (DOI, 1999).

BOX 1-1
HAZARDS IN THE PACIFIC NORTHWEST

The Cascadia project of the CMGP has acquired three major data sets that address regional crustal structure of the Cascadia margin across (1) the Mendocino Triple Junction, (2) the Washington continental margin, and (3) the Puget Sound lowland (USGS, 1998a). These projects, in conjunction with the USGS Earthquake Hazards Program, have revealed a number of active faults that have large seismic potential on the margin and in the Seattle-Tacoma region. These faults, which are near the earth's surface, pose as great a threat to the inhabitants of this region as the more distant subduction-zone earthquakes. For this reason, earthquake hazard assessments have focused on mapping the onshore and offshore distribution of faults, assessing how the faults are evolving, and understanding how the upper plate structures are linked to subduction at depth (Fluck et al., 1997). Ground motion hazard maps developed at the USGS are integrally linked to the proximity of these faults, the likely recurrence of earthquakes on these faults, and the wave velocity of the intervening blocks.

Efforts fielded to obtain the three data sets have involved consortia of government and academic institutions, coordinated and managed through the USGS. The most recent experiment to image the Puget Sound region involved collaboration between the USGS and the Geological Survey of Canada, including the contracting of vessels for both surveys. Onshore participants from the USGS, Oregon State University, and the University of Washington deployed seismometers throughout the Puget Sound lowland to enhance refraction models. The prior study involved cooperatives with Forschungszentrum fuer Marine Geowissenschaften (GEOMAR) in Kiel, Germany, the University of Washington, and Oregon State University. The program initiated contacts and coordinated the various entities participating in the project. Roles were determined largely on scientific expertise. The CMGP staff took the lead in coordinating the marine component of the experiments, whereas responsibilities for coordinating the onshore work were equally shared by CMGP, other USGS participants, and university participants.

Seattle's Project Impact is a multidisciplinary assessment of the geologic hazards affecting the Seattle-Tacoma corridor, with the direct purpose of developing mitigation efforts (University of Washington Geophysics Program, 1998). This project is incorporating the results of the experiments and in particular (1) the newly revealed style structural of faults in the Puget Sound region and (2) the velocity structure of the Puget Sound region. The character and proximity of the faults and the velocity structure directly determine ground motion potential and, therefore, the potential impacts of motions on the built environment.

Project Impact is a collaborative effort between state agencies, federal agencies, the engineering community, and the private sector to identify the critical weaknesses in and to improve infrastructure exposed to earthquakes and to improve infrastructure performance. Toward that end, the data and interpretations developed at CMGP are essential.

by sediment input and has been key to meeting the energy needs of our nation with its large reservoirs of oil and gas. Hence, the nature of the offshore environment and the geologic processes at work are inextricably linked to the geologic framework of the near-shore regions and their resources and ecosystems.

As an integral part of the nation's assets, it is critical that the health and viability of the coastal and marine environment be preserved. To manage these important resources responsibly, the complex natural systems that comprise these regions must be characterized and ultimately modeled. An understanding of a key factor that provides the underpinning of any useful coastal and marine system model is the geologic framework on a variety of scales (e.g., from the scale of its place in global plate tectonics, to the geologic formations resulting from local geologic processes, to the movement of sand grains on a beach). A focused effort of scientific study and national assessment, directed toward developing an understanding of the geologic framework, of this vast, complex, and interconnected coastal and marine region, is necessary if the nation is to wisely manage this important asset.

THE VALUE OF UNDERSTANDING GEOLOGIC PROCESSES

Geologic processes control the fundamental nature of the earth's surface. The fundamental character of the margins of the world's continents are the result of interactions of many tectonic plates that form the earth's crust. This character is then modified through alternating periods of sediment erosion and deposition, and additional crustal movement. Thus, the character of the U.S. coast and continental margins reflects the delicate interplay of a number of natural processes, as well as the footprint of human activity. The size, depth, and shape of coastal rivers, estuaries, and beaches reflect the control of these same geologic processes through time. Geologic processes also control the size, shape, volume, and quality of freshwater aquifers along the coast. Furthermore, the distribution of wetlands, fishing grounds, minerals, sand and gravel, and other natural resources are also impacted by these processes. Thus, the key to wise stewardship and development of many coastal and marine natural resources is a sound scientific understanding of how earth systems operate. Science-based decisions facing federal, state, and local policymakers can be expected to depend on an understanding of the processes that have traditionally been the focus of research by the USGS and its CMGP.

ROLE OF THE USGS

The USGS was established by the Organic Act of March 3, 1879 (20 Stat. 394; 43 U.S.C. 31), which provided for "the classification of the public lands and examination of the geologic structure, mineral resources, and products of the national domain." In 1997, the Biological Resources Division was added, bringing another disciplinary component into the agency's realm of responsibility. Thus, the mission of the USGS is to provide biologic, geologic, topographic, and hydrologic information that contributes to the wise management of the nation's natural resources and that promotes the health, safety, and well-being of the people. This information consists of maps, databases, and descriptions and analyses of the water, energy and mineral resources, land surface, biologic resources, underlying geologic structure, and dynamic processes of the earth's ecosystems. To accomplish its mission, the USGS:

- conducts and sponsors research in geology, mapping, hydrology, and related sciences; describes the onshore and offshore geologic framework and develops an understanding of its formation and evolution; assesses energy and mineral resources, determines their origin and manner of occurrence, and develops techniques for their discovery; evaluates hazards associated with earthquakes, volcanoes, floods, droughts, toxic material, landslides, subsidence, and other ground failures, and develops methods for hazard prediction; produces and updates geographic, cartographic, and remotely sensed information in graphic and digital form, and develops advanced mapping techniques, as well as new applications for cartographic and geographic data; collects and analyzes data on the quantity and quality of surface water and groundwater, on water use, and on quality of precipitation; and assesses water resources and develops an understanding of the impact of human activities and natural phenomena on hydrologic systems.
- publishes reports and maps, establishes and maintains earth science databases, and disseminates earth science data and information; provides scientific and technical assistance in the effective use of earth science techniques, products, and information; and develops new technologies for the collection, coordination, and interpretation of earth science data.
- coordinates topographic, geologic, and land-use mapping, digital cartography, and water data activities in support of national needs and priorities; provides scientific support and technical advice for legislative, regulatory, and management decisions; and cooperates with other federal, state, and local agencies and with academia and industry in the furtherance of its mission.

THE GEOLOGIC DIVISION

The USGS is organized into four divisions: Water Resources Division, Biological Resources Division, National Mapping Division, and Geologic Division. The USGS Geologic Division (GD) is the nation's primary federal provider of objective, relevant, and reliable earth science information on geologic hazards, energy and mineral resources, geologic framework, and coastal and marine processes. To provide this information, the division conducts geological, geophysical, and geochemical surveys and investigations throughout the United States, its island territories, and its EEZ and cooperates in global geophysical monitoring and foreign disaster assessments. It conducts geologic mapping to establish the composition, structure, and geologic history of sediments and rocks at and beneath the earth's surface. These maps convey information critical to understanding and assessing the endowment of U.S. mineral and energy resources, to maintaining the environmental quality of lands and waters, and to understanding and mitigating the effects of geologic hazards. The GD coordinates closely with a broad constituency of federal (including other USGS divisions), state, and local agencies; other public and private sector entities; and international agencies and institutions to assure that these agencies' earth science information needs are identified and met in a timely manner. The information is made available in electronic and printed form as assessments, interpretative reports, maps, and data and through personal communications, such as workshops, forums, and meetings.

To accomplish this mission, the GD conducts surveys, investigations, and research on:

- such extreme natural events as earthquakes, volcanic eruptions, landslides, subsidence, erosion from coastal storms and hurricanes and geomagnetic storms that inflict an average annual loss of scores of lives and billions of dollars of damage. Division activities provide information and data with which to make informed management and policy decisions that reduce economic risks and improve public safety.
- natural geologic processes and phenomena and human-induced actions that operate at the earth's surface and control the evolution of landscapes and the resulting quality of the physical environment.
- the location, quantity, quality, and availability of mineral and energy resources.
- the economic cycle of minerals, including production, consumption, recycling, stock, and shipments for some 100 commodities and 190 countries.
- the geologic and geophysical framework and natural processes that shape U.S. land and coastal and marine areas and thereby affect human activities.
- the global environmental systems and past and present climate changes in order to document the natural variability of the climate system, establish the environmental consequences of past climate change and likely future climate change for sensitive regions, and monitor related physical properties that could indicate changing environmental conditions.
- the physical framework and processes of planetary bodies.

The field organization of the GD consists of three regional offices, each led by a regional geologist. Regional geologists represent the interests of the chief geologist in furthering the objectives, policies, and procedures of the division. They are responsible for implementing scientific program activities of the division and developing and managing inter-division programs and projects and most reimbursable work supported by other federal and state agencies. They exercise

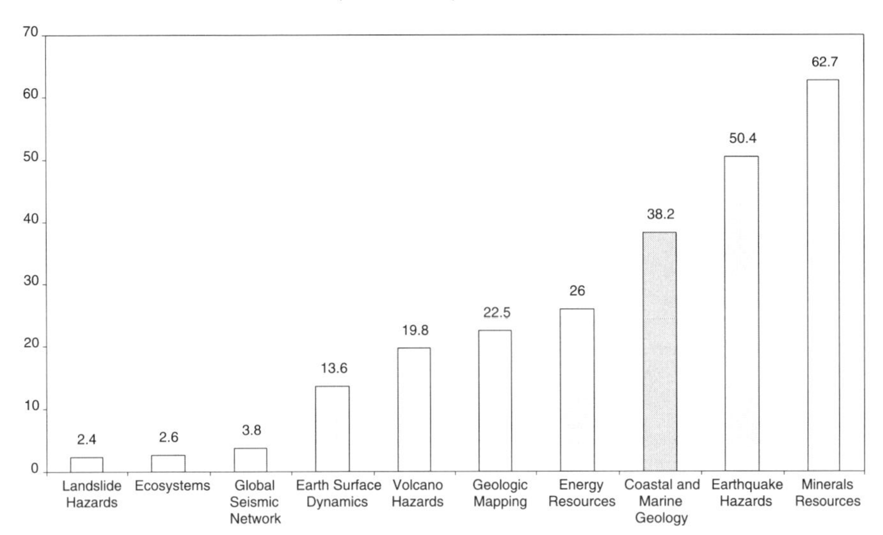

FIGURE 1-2 Comparison of program budgets (FY 1999) for the 10 programs of the Geologic Division of the USGS (Appendix E).

line management authority in their particular region, which is the level at which division programmatic activities are carried out.

The GD provides scientific and technical assistance to other federal, state, and local agencies; public and private sector entities; and international agencies and institutions requiring geological, geophysical, or geochemical information and assessments. The work of the GD is carried out largely through a series of coordinated activities (Fig. 1-2) that include the Energy Program, the Minerals Program, the Earthquake Program, the Volcano Hazards Program, and the CMGP. As shown in Figure 1-3, the CMGP represents a relatively small portion of the overall effort of the USGS, when measured in terms of financial expenditures. However, the CMGP represents the focusing mechanism for bringing the considerable scientific capabilities of the USGS to bear on obstacles limiting the nation's ability to manage its coastal and marine resources wisely.

THE COASTAL AND MARINE GEOLOGY PROGRAM

The CMGP allows the scientific expertise in the GD to be applied to a range of issues that have broad policy implications for the U.S. Department of the Interior (DOI), the federal government, and the nation as a whole (e.g., coastal and offshore hazards, marine resources development and conservation, and effect of climate change). In the federal government, many agencies are involved in

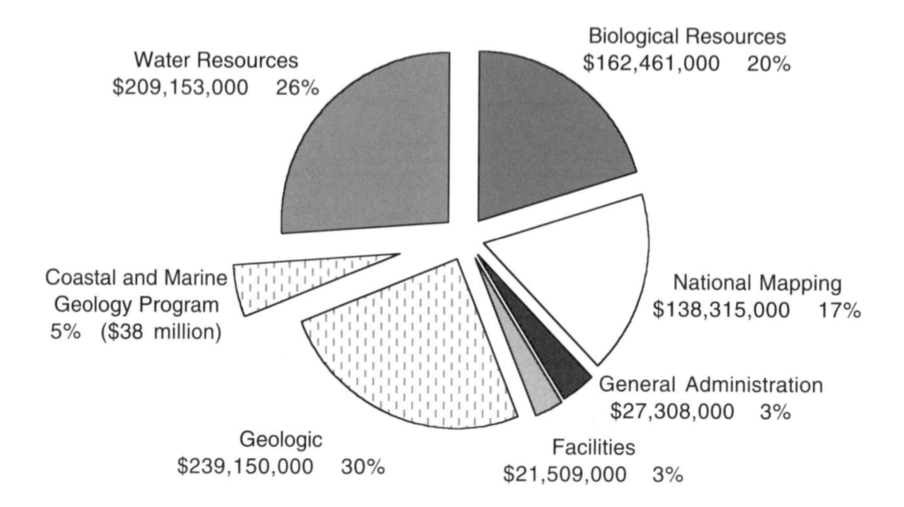

FIGURE 1-3 The Coastal and Marine Geology Program as a percent of the FY 1999 USGS budget (Appendix E).

activities along the continental margin (U.S. Environmental Protection Agency [EPA], U.S. Army Corps of Engineers [USACE], National Oceanic and Atmospheric Administration [NOAA], the Navy, etc.), but the USGS's CMGP is the only program whose mission is to establish the geologic framework of coastal regions; it is also the only program that has the scientific resources needed to place the complex interactions in the coastal region, which operate over a wide range of spatial and temporal scales, into a geologic context. The program's scientific resources are currently distributed among three regional centers.

The CMGP addresses issues of national importance in the areas of environmental quality and preservation; natural hazards and public safety; and natural resources, providing information on coastal and marine geology for the science community and public benefit. The program provides information and products to guide the preservation and sustainable development of the nation's coastal and marine environment, including both the EEZ and the Great Lakes (Box 1-2).

CMGP's research and mapping investigations are designed to describe coastal and marine systems; to understand the fundamental geologic processes that create, modify, and maintain them; and to develop predictive models.

INTENT OF THIS STUDY

In June 1994, the USGS implemented a National Coastal and Marine Geology Plan outlining proposed studies and budgets for understanding the coastal

BOX 1-2
LAKE ERIE COASTAL EROSION STUDY

Erosion of the Ohio shoreline of Lake Erie was recognized as a threat to private property and public infrastructure. The State of Ohio, mandating that coastal erosion areas be identified in development of Ohio's Coastal Management Program, discovered a pressing need for reliable information documenting ongoing erosion rates. Ohio law mandates that coastal erosion areas encompass those "areas anticipated to be lost over the next 30 years due to Lake Erie-related erosion if no additional coastal erosion control measures are emplaced." To meet this requirement a cooperative program was designed to improve estimates of ongoing erosion and to provide a regional understanding of the factors involved (Folger, 1996; Mackey, 1996).

The effort involved the USGS and the Ohio Department of Natural Resources' and the Ohio Division of Geologic Survey (ODGS). ODGS provided both scientific expertise and regulatory oversight for assessing development and management in Ohio's coastal erosion areas. The USGS provided scientific expertise and technical capabilities for data collection and interpretation. The bulk of the USGS expertise came from the CMGP, which was also responsible for project oversight. Additional expertise was supplied by the USGS National Cooperative Geologic Mapping Program.

As the agency responsible for erosion control and issuing development permits in the resulting coastal erosion areas, the ODGS was largely responsible for identifying the data and product needs for the specific regulatory requirements. The ODGS and the USGS cooperatively defined a data collection and analysis program to meet these immediate needs while enhancing understanding of the processes driving shoreline retreat. Efforts centered on documentation of recession rates and factors (bluff lithology, shoreline modification, sediment entrapment) contributing to shoreline retreat. A broad regional approach was cooperatively designed and implemented to address the role of restricted sand resources on the future evolution of the shoreline. A team of ODGS and USGS research and technical staff cooperatively implemented the program. The ODGS took responsibility for dissemination of results to the public and the ultimate development of policy. The USGS was responsible for ensuring the quality and defensibility of the scientific interpretation. Both agencies participated fully in the data collection and interpretation programs. The interaction between the two agencies was highly successful in the definition of program objectives that addressed both scientific and management needs.

The study resulted in the establishment and approval of a Coastal Management Program for Ohio. Designation of coastal erosion areas reflected the study findings that long-term shoreline retreat rates had been significantly impacted by fundamental changes in the nearshore system. A progressive, dramatic reduction in beach width and sediment supply since the early 1970s, due in large part to the emplacement of shore protection structures and high lake levels, had caused an acceleration of erosion along the unprotected areas of the coast. Based on data resulting from this study and recognizing a system-wide change in coastal conditions, the state modified administrative rules to use short-term recession rates (maximum 30-year interval) to designate coastal erosion areas.

BOX 1-3
STATEMENT OF TASK

The committee will review the history and status of the Coastal and Marine Geology Program, particularly its most recent national plan and recent workshop reports in the context of the U.S. Geological Survey and the new Geologic Division's Science Strategy. The committee will review current studies and science staff by means of visits to the research centers in Woods Hole, St. Petersburg, and Menlo Park.

The committee will then provide advice on:

- the general areas of future program emphasis (e.g., research, national assessments, monitoring, characterization) and cooperation with local, state, and national decisionmakers, and with government and academic scientists;
- the specific scientific and technical challenges (including components from the national plan such as coastal erosion, earthquake hazards, pollution studies, biologic habitats, distribution and significance of gas hydrates), as well as the challenge to maintain a strong and dedicated research staff;
- balancing between issue-driven and knowledge-driven research, and between regional and national efforts;
- the ideal mix of science staff as to discipline and status (permanent versus term) to meet needs and ensure the long-term health of the Coastal and Marine Geology Program; and
- the ideal ratio of core-funded research versus reimbursable research paid by clients.

and offshore areas of the United States and its territories. Since the plan was accepted by Congress several changes have occurred that affect the program: (1) departmental budgets have been level; (2) funding from outside sources has been modest; (3) the GD underwent a significant (25 percent) downsizing in staff; (4) the GD was reorganized, and the former Office of Marine Geology was renamed the Coastal and Marine Program to reflect the fact that the programmatic emphasis has shifted from deep water to the shelf, coast, and estuaries. In addition, many new issues and opportunities unforeseen in the original plan have arisen. These changes made it appropriate to outline new directions for the CMGP in a revised national plan released in 1997 (USGS, 1997).

As part of an ongoing GD effort to receive input from the broader scientific community, the GD requested that the National Research Council (NRC) conduct reviews of a number of its ongoing programs, including the CMGP. In response to this request the NRC formed the Committee to Review the USGS Coastal and Marine Geology Program (Box 1-3). Early in its deliberations, the committee recognized that its review of the CMGP would be held against a backdrop of planning activities that have taken place across the USGS. Both the USGS as a whole and the GD have undertaken efforts over the last two years to develop strategic plans. The committee therefore recognized that, in advising the CMGP, it would need to consider the concepts outlined in those plans.

For example, in the fall of 1997, the USGS developed a strategic plan to guide the organization into the 21st century. The committee found much of the logic and goals contained in the plan to be both thoughtful and prudent; thus, much of the advice given in this report is in keeping with many of its basic tenets. Of particular value is a brief discussion of the vision of the USGS for 2005 (as articulated in the strategic plan).

> The challenge for the USGS is to stay focused on a horizon of some ten years out, while realizing that there will be near-term shifts that will demand our scrutiny and perhaps mid-course corrections. These shifts and corrections will be driven by such forces as the increasing devolution of federal government functions to the states and other entities, changes in national demographics, the expanding influence of advances in scientific methods and technologies, and the continuing—and underlying—tension between the development of the nation's natural resources and environmental conservation. Beyond these already compelling factors are the public's perception of its investment in science as a means of solving societal problems and society's concept of the "public good" of science. . . .
>
> What will characterize the U.S. Geological Survey in 2005? The USGS will be focused on a well-defined group of business activities. The level of effort applied to current activities will be different. For example, the USGS will conduct more studies on hazards, water, and contaminated environments and fewer studies on non-renewable resources.

The following are the salient changes in emphasis mentioned in the USGS strategic plan:

Increasing Emphasis

- long-term interdisciplinary studies
- mitigation studies
- quality and accessibility of resources
- international mineral/energy studies
- nontraditional disciplines
- regional and national studies
- geospatial data integration
- applied research and development
- technology transfer
- issue-driven studies
- studies involving population centers
- multiple-risk assessments
- digital products
- real-time event responses

Decreasing Emphasis

- single-discipline studies
- remediation studies
- distribution and quantity of resources
- domestic mineral and energy studies
- traditional earth science disciplines
- local studies
- sole production of geospatial data
- basic research studies
- compartmentalized technology
- investigator-driven studies
- wilderness areas studies
- single-risk assessments
- paper products
- post-event responses

Similar concepts from both the USGS strategic plan (USGS, 1997) and the GD's strategic plan (USGS, 1998h) appear throughout this report to acknowledge the committee's recognition that these documents provide particularly helpful guidance to the CMGP.

SCOPE OF THIS REPORT

As outlined in its statement of task, the committee collected a great deal of information during its visits to each of the three research centers and through extensive input from the USGS staff and many of the users and collaborators of the program. In total the committee either heard presentations or received written input from individuals representing 25 federal, state, and local government agencies, academic institutions, and nongovernmental groups. Those discussions centered on the policy decisions facing entities responsible for coastal and marine areas in the United States and the role the CMGP plays or should play in providing necessary information to support science-based decisionmaking.

This report is organized according to the committee's charge and is intended for multiple audiences, including scientists familiar with both coastal and marine geology and the CMGP, and policymakers who may not be familiar with either. Chapter 2 discusses the current niche, as understood by the committee, of the CMGP in addressing research, assessment, monitoring, and characterization of U.S. coastal and marine areas. Chapter 3 speaks to the overarching or grand challenges facing these areas, and Chapter 4 sketches the central role the committee feels the CMGP should play in addressing those challenges, as well as more specific near-term scientific and technical challenges. Chapter 5 contains the committee's advice on how critical human and technical resources can be focused to allow the CMGP to implement change successfully and position itself to play a leading role in advancing this nation's ability to manage its coastal and marine resources and promote the health, safety, and well-being of the people.

2

The Coastal and Marine Geology Program

The Coastal and Marine Geology Program (CMGP) is an extremely important component of the Geologic Division (GD) of the U.S. Geological Survey (USGS). It conducts research and provides data on the critical interface between land and sea and on the continental margins out to the limit of the U.S. Exclusive Economic Zone (EEZ). It is currently the third-largest program (in terms of funding) in the GD (after Mineral Resources and Earthquake Hazards), with a FY99 appropriation of $38.2 million (Figs. 2-1a and 2-1b).

The role of the USGS in conducting studies to understand the coastal and marine areas of the United States was acknowledged in 1994 with the implementation of a five-year National Marine and Coastal Geology Program Plan (USGS, 1994b). This plan was then modified in 1997 to take advantage of new opportunities and issues and to account for changes in budgets and staffing (USGS, 1997).

The stated mission of CMGP is to "provide the nation with objective and credible marine geologic science information based on research, long-term monitoring, and assessments." CMGP is designed to describe marine and coastal geologic systems; to understand the fundamental processes that create, modify and maintain them; and to develop the capability to predict future change through models that integrate the characteristics of natural systems and the effects of human activities.

In the committee's opinion, CMGP conducts unbiased, high-quality scientific research and provides key geologic data and information to address issues along the U.S. coast and within the U.S. EEZ. The broad base of CMGP scientific and technical expertise allows the creation of diverse teams that can (i) conduct integrated, field-based scientific research in all coastal and marine envi-

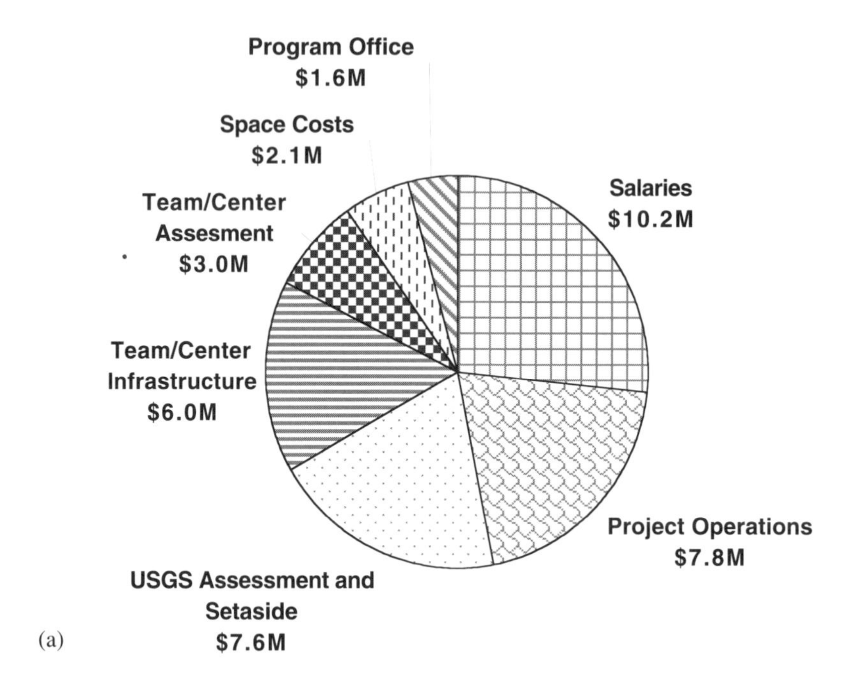

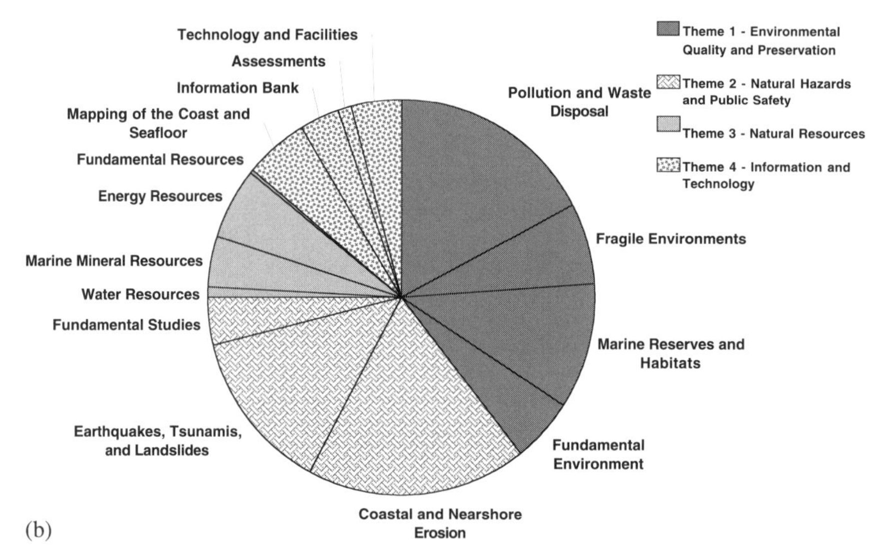

FIGURE 2-1 (a) Breakdown of the Coastal and Marine Geology Program budget (FY 1999) (Appendix E). (b) Breakdown of funds used to support specific research efforts (including staff salary and operating expenses, FY 1999) by theme (Appendix E).

ronments; (ii) investigate and model interactions among geologic, chemical, and fluid processes; and (iii) complete large, long-term, regional and national, and multidisciplinary studies and assessments of coastal and marine geologic issues. This view of the role of the CMGP was supported by perspectives provided by USGS staff, federal and state agencies, and other users and collaborators (Appendix C).

THEMES OF THE CMGP

Effective guidance for CMGP's future must be predicated on a solid understanding of the nature and abilities of the CMGP today. Consequently, the committee spent considerable time reviewing ongoing projects and present capabilities of the CMGP. These are discussed here to more fully enlighten the reader about CMGP as it exists today and to form a basis for future change.

The 1997 Five-Year Plan identified four themes as the focus of investigations in CMGP: 1) environmental quality and preservation, 2) natural hazards and public safety, 3) natural resources, and 4) information and technology. Studies in each of these scientific themes are broken down into two general types: fundamental and regional. Fundamental studies, which typically account for approximately 5 percent of the annual CMGP budget (Fig. 2-1b), are designed to improve the basic quantitative understanding of the complex geologic processes active in the marine and coastal environments. Results from such studies are relevant to a wide variety of coastal and marine regions, and they also enhance predictive capabilities useful for anticipating future long- and short-term changes. Regional studies typically develop a description of a specific marine and coastal geologic system where problems significant to specific subthemes are identified. A description of the present CMGP focus and activities, organized by the themes and subthemes, is presented below.

Theme 1: Environmental Quality and Preservation

With the growing pressures from human activities along the U.S. seaboard, the quality and preservation of the coastal and marine environment have become urgent issues. Science-based management of these areas requires the development of a basic understanding of the natural and anthropogenic factors that influence the quality of the environment. The CMGP has a role in investigating the dynamics of geologic processes affecting our coastal and marine environments today through sampling, data collection, and modeling (e.g., the Large-Scale Coastal Modeling Project and the Inner Shelf Dynamics Project). In addition, by participating in such multiagency, multiinstitutional projects as STRATAFORM (the Origin of Marine Stratification), the CMGP also examines long-term geologic changes through studies of the sedimentary records preserved in seafloor environments. Such studies provide the geologic framework of the coastal areas and continental margins and define the geologic processes that underpin many factors impacting the quality of the marine environment.

The other major area of concern is sea-level change. There is a pressing need to develop models to predict the future of sea-level change and its impact on the United States. Areas particularly sensitive to sea-level changes are the coastlines, coastal wetlands, and coral reefs. The CMGP is conducting regional studies to account for both global sea-level changes and local land subsidence and uplift along the coastline. An excellent example of such a study is the investigation of the effects of sea-level rise and subsidence on the Louisiana coast.

Subtheme 1: Pollution and Waste Disposal

The legacy of the use of the ocean for waste disposal and the present and future management of such activities are issues of considerable concern. The range of waste materials and pollutants (e.g., heavy metals, garbage, radioactive waste, nutrients, organic chemicals, and microbes), and their variety of behaviors in the coastal and marine environment, require that a combination of basic scientific and applied regional studies be undertaken (see Box 2-1).

The CMGP is studying sediment transport processes and the long-term fate of pollutants (e.g., in Massachusetts Bay, Long Island Sound, Florida's Santa Monica Bay, Hawaii, Monterey Bay, Lake Pontchartrain); the physics, chemistry and biology of sediment-pollutant interactions (Boston Harbor); and sediment mixing and sorting processes (e.g., mechanisms of sorting in sand and shell hash beds on the west Florida shelf) to better understand the controls on the distribution and ultimate fate of waste materials and pollutants in the ocean. Regional studies are also under way to map the distribution of contaminants (e.g., barrels of radioactive waste in the Gulf of the Farallones; hydrocarbons in Prince William Sound); to model local circulation patterns and the associated sediment transport (e.g., on the Los Angeles shelf and on the continental slope off San Francisco); and to investigate the local processes that could result in remobilization of contaminants from the seafloor (e.g., metal concentrations in Boston Harbor sediments).

Subtheme 2: Fragile Environments

Coastal and marine environments are host to a variety of fragile environments that exist in areas of transition where a delicate balance between two types of environments must be maintained. For example, the coastal wetlands that act as the nursery grounds for species comprising about 80 percent of U.S. commercial and recreational fisheries occur at the interface of salt- and freshwater and require a delicate balance among physical and chemical oceanographic processes to maintain their health. Along the continental slopes and on the mid-ocean

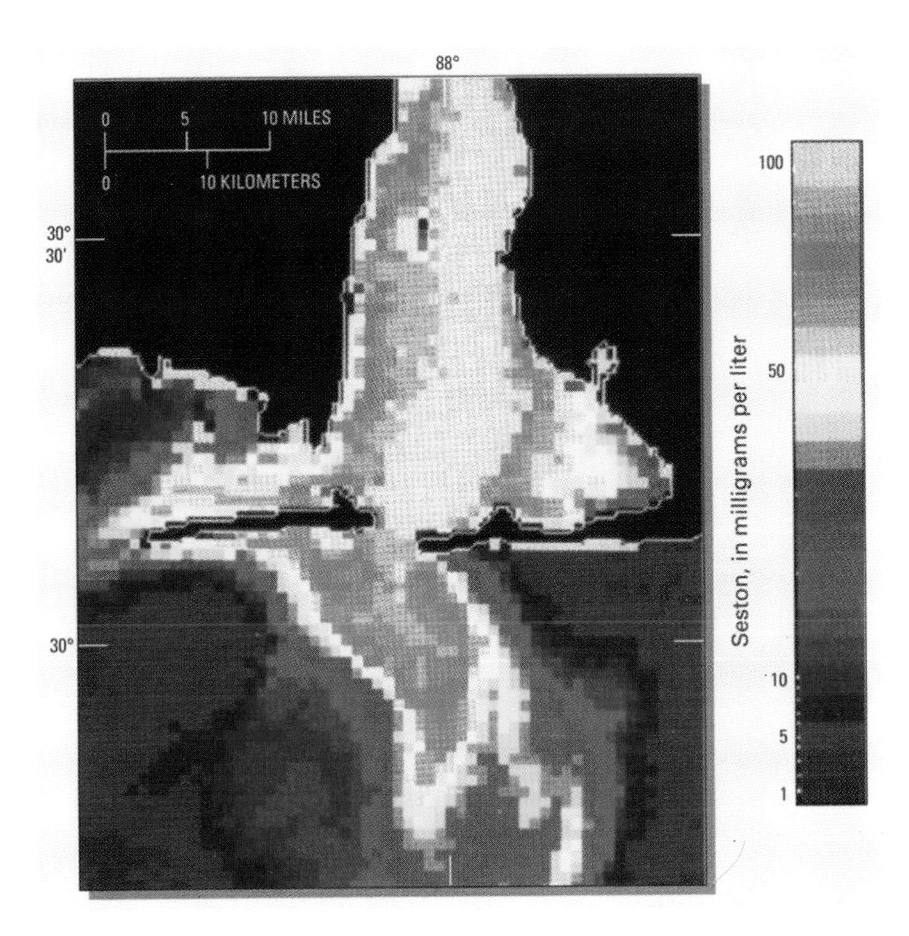

PLATE 1 Satellite image of suspended sediment concentration showing a large plume extending 60 kilometers offshore during high river flow. The silt-laden river water flows through Mobile Bay, forming the plume as it floats on the saltier and denser Gulf water (USGS, 1994a).

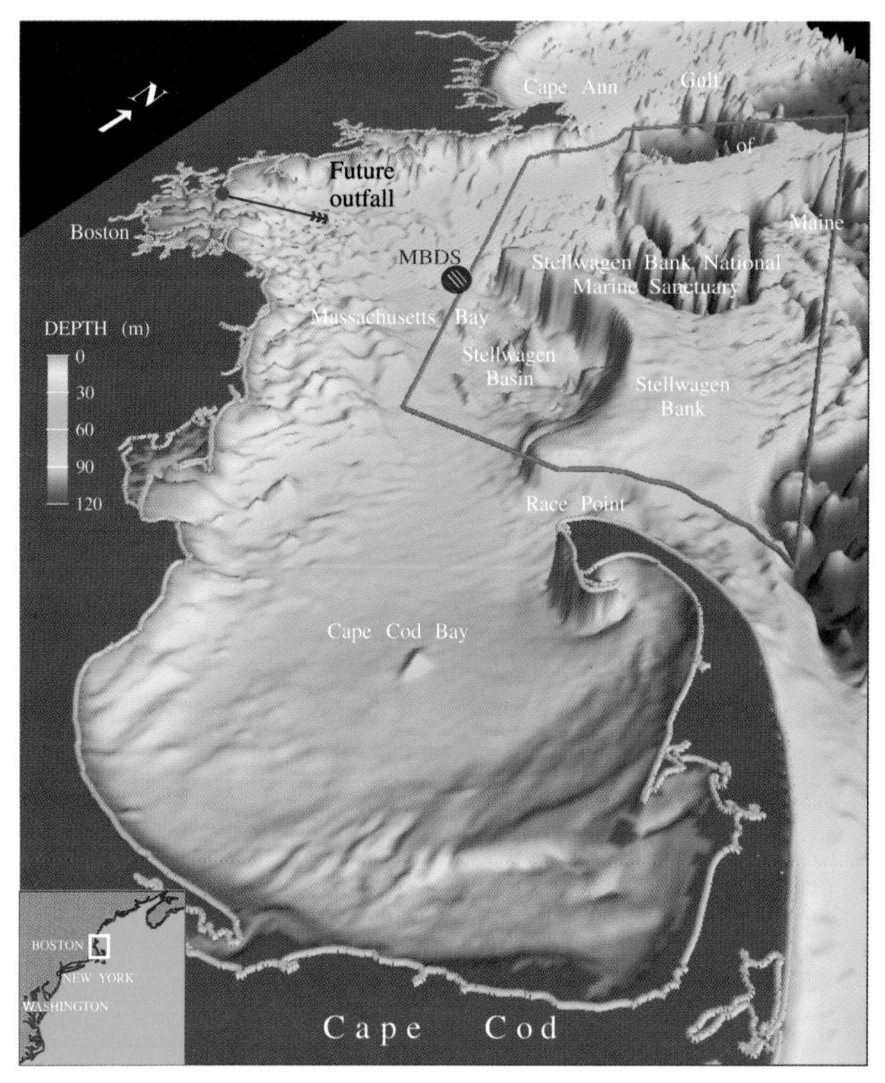

PLATE 2 Perspective map of Massachusetts Bay and Cape Cod Bay illustrating the complex underwater topography. The region is approximately 100 km long and 40 km wide. Stellwagen Bank rises to within about 20 m of the sea surface and partially isolates Massachusetts Bay from the Gulf of Maine. Beginning in 2000, the discharge of treated sewage effluent from the Boston metropolitan area will be relocated from Boston Harbor to a new site 15 km offshore (approximately 35-m water depth) in Massachusetts Bay. The location of the Deer Island treatment plant (red dot), the future outfall, the Massachusetts Bay disposal site (MBDS), and the Stellwagen Bank National Marine Sanctuary (SBNMS) are also shown. Note that the MBDS is located outside the SBNMS. Vertical exaggeration is 100X (USGS, 1998b).

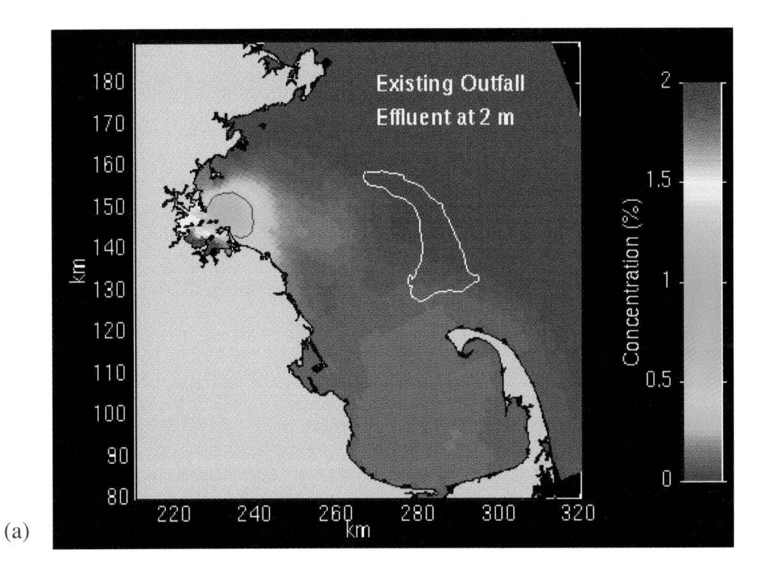

(a)

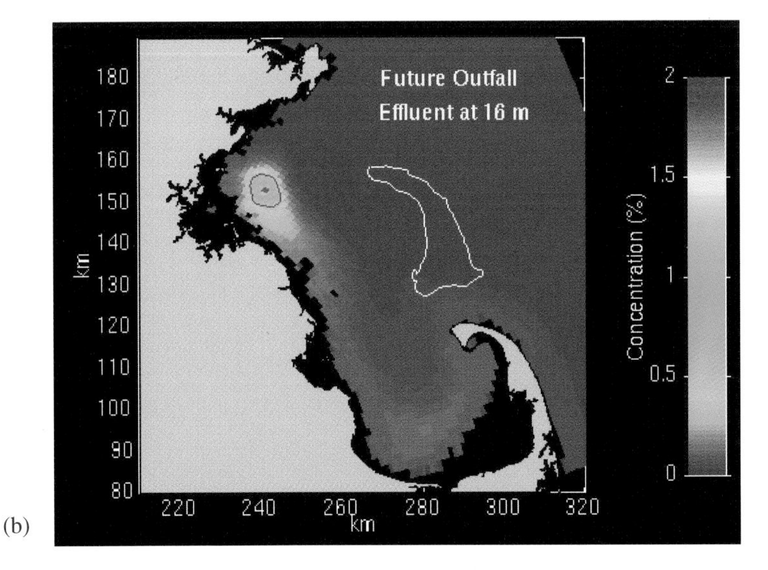

(b)

PLATE 3 Model comparison of (a) summer near-surface (2 m depth) effluent concentrations off of Boston Harbor at the existing sewage outfalls and of (b) summer middepth (16 m depth) concentrations at the new outfall. At the new outfall location, effluent is trapped at mid-depth during the summer beneath the warm surface layer, while effluent from the existing outfalls remains near the surface. The areal extent of high effluent concentration at the new outfall is smaller, as in winter, than at the existing outfalls. In addition, because nutrients from the new outfall are trapped in waters that are already nutrient rich, the impact of sewage-borne nutrients is decreased (USGS, 1998c).

(a)

(b)

PLATE 4 (a-b) (a) Typical healthy coral ecosystem prior to 1983. Coral reefs are built by tiny coral polyps, simple soft-bodied animals living in colonies. Each polyp constructs an outer skeleton of hard limestone that adds a new layer to the reef. Spiny sea urchins *Diadema antillarum* grazing green algae from dead coral surfaces (white areas). Beige areas are live coral polyps. (b) Black-band disease on a brain coral in a reef at Rum Cay in the eastern Bahamas in 1990. This and other types of band diseases can destroy a 200-year-old head coral in a single summer.

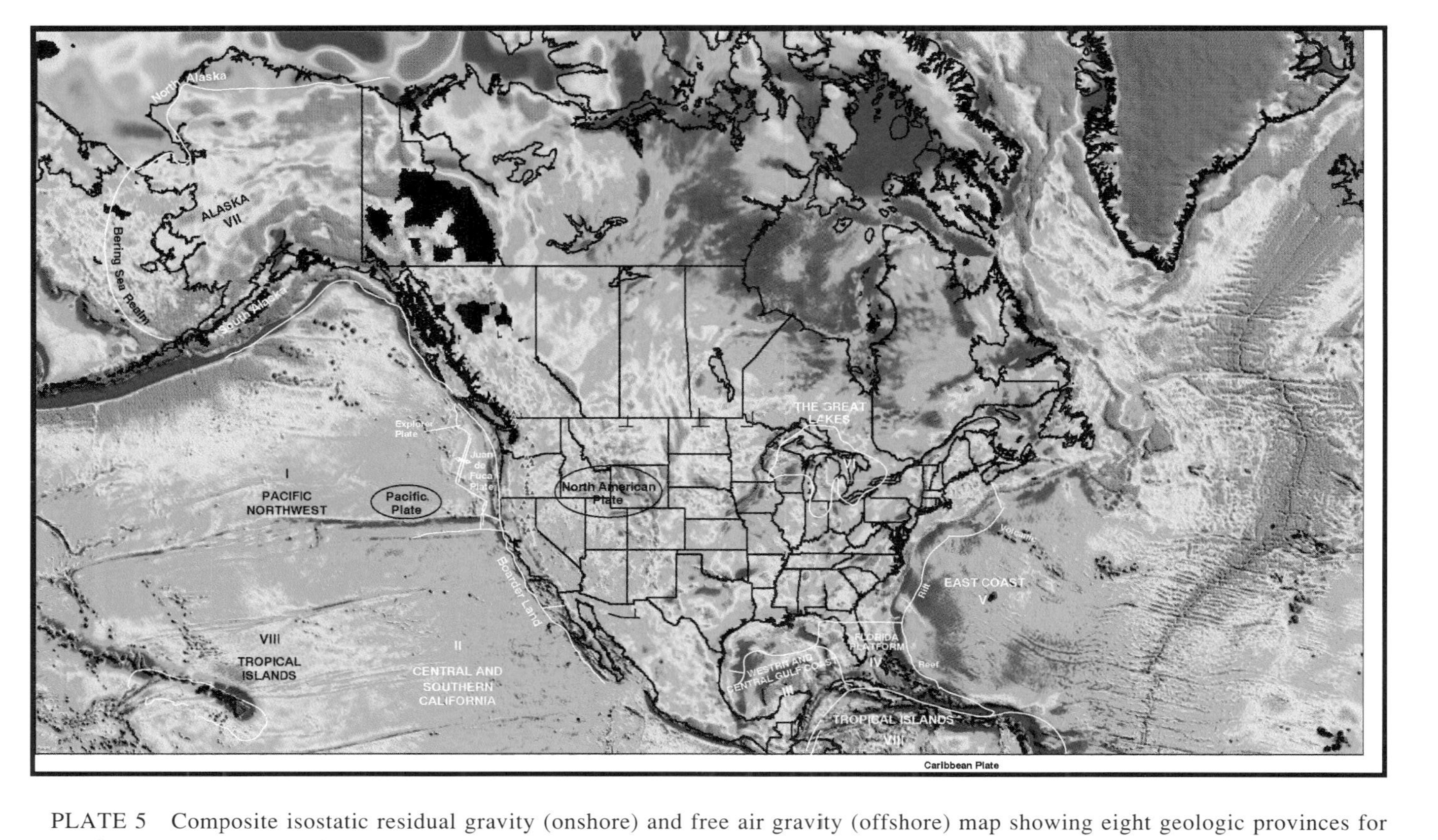

PLATE 5 Composite isostatic residual gravity (onshore) and free air gravity (offshore) map showing eight geologic provinces for North America (modified from Geological Society of America 1988; Sandwell and W. Smith 1997).

(a)

(b)

PLATE 6 West Onslow Beach, N.C. overflow (34º 30.70'N, 77º 22.16'W). (a) After Hurricane Bertha, 7/16/96, 10:34 EDT. (b) After Hurricane Fran, 9/7/96, 14:38 EDT. Areas overwashed by Hurricane Bertha were more susceptible to erosion and overwash by Hurricane Fran. Initially susceptible to erosion during a category 2 hurricane, the area affected by the category 3 storm was greatly expanded. (Stumpf et al., 1996).

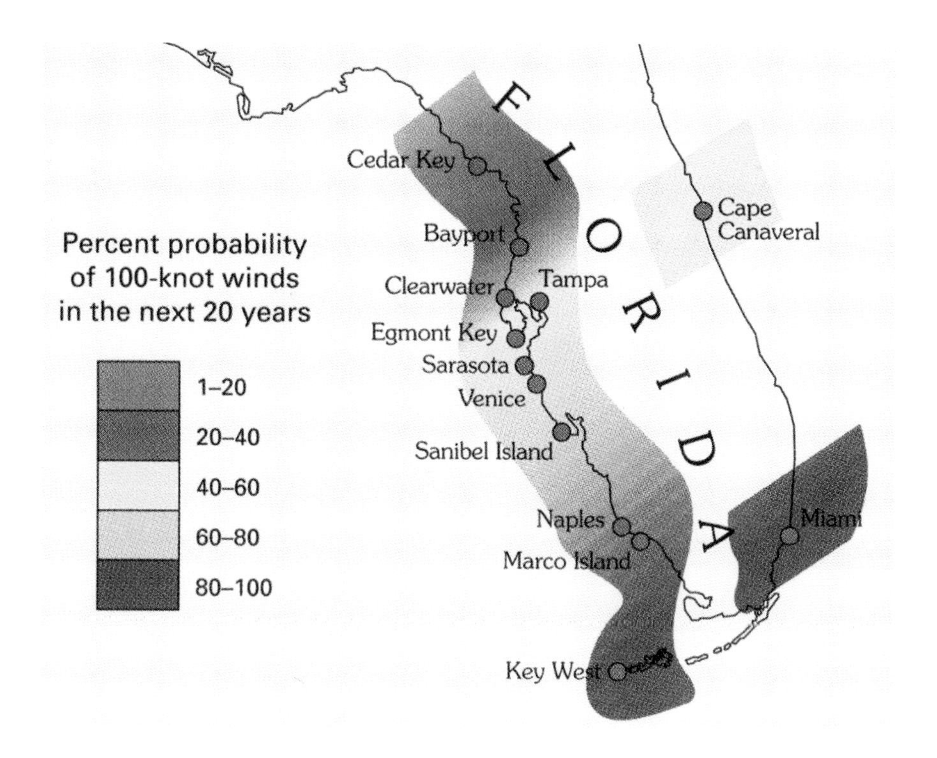

(a)

PLATE 7 (a) The frequency of Florida hurricanes with wind speeds greater than or equal to 100 knots is mapped in terms of the probability of occurrence during a 20-year exposure window. These probabilistic estimates, based on 106 years of observations, illustrate that hurricanes with 100-knot winds occur more frequently in southern Florida and gradually decrease in frequency towards northern Florida. (b; next page) Map showing tracks of the deadliest and costliest hurricanes in the United States between 1900 and 1993 (USGS, 1994a).

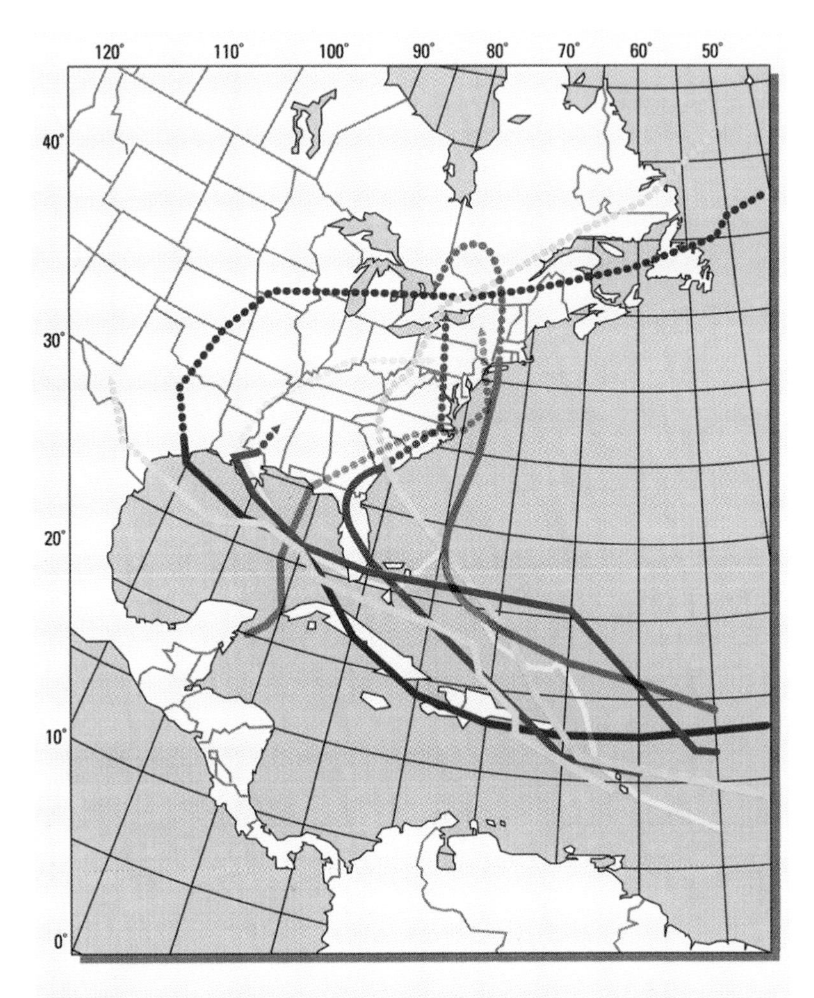

DEADLIEST	Land fall	Lives lost
Galveston	Sept. 8, 1900	6000
Lake Okeechobee	Sept. 17, 1928	1836
New England	Sept. 21, 1938	600
Florida Keys and Southwest Texas	Sept. 14–16, 1919	600

COSTLIEST	Land fall	Billions of U.S. dollars*
Andrew	Aug. 24–26, 1992	25.0
Hugo	Sept. 22, 1989	7.2
Betsy	Aug. 10, 1965	6.5
Agnes	June 22, 1972	6.4

* (1990 value)

(b)

BOX 2-1
BOSTON HARBOR

The results of the USGS study of Boston Harbor and Massachusetts Bay have been used to help make management decisions throughout the Boston Harbor Cleanup Program. The focus of USGS activities is to provide an understanding of the fate and transport of contaminated sediments (USGS, 1998b, 1998c, 1998d, 1998h, 1999).

The USGS side scan sonar maps of the seafloor in western Massachusetts Bay (Plate 2) were used by the Massachusetts Water Resources Authority (MWRA) to help decide between two alternative sites for Boston's new sewage outfall (MWRA, 1996). According to Paul F. Levy, former executive director of MWRA, the timely production of this map saved MWRA the significant expense of geotechnical studies of the rejected site.

USGS maps of the bottom characteristics in Boston Harbor, Massachusetts Bay, and Cape Cod Bay were used in the design of the federally required MWRA monitoring program. The maps contributed to a more efficient and cost-effective monitoring program by identifying areas of similar sediment types (that could then be characterized by fewer measurements), as well as areas where changes over time could be most clearly documented. The maps of bottom types also have been useful to fishermen.

USGS computer models of circulation (Fig. 2-2) illustrate the comparative impact of sewage from Boston's existing and future outfall (Plate 3). The models suggest that, when the outfall is moved to the offshore location, water quality will improve dramatically in Boston Harbor and neither the beaches of Cape Cod nor the area around Stellwagen Bank will be exposed to increased sewage. The model was used by the U.S. Attorney's Office (Department of Justice) in defending the government in the endangered species case concerning right whales in the Stellwagen Bank National Marine Sanctuary. The results also were useful in public and congressional hearings by providing an unbiased assessment of the consequences of management actions.

The model results also helped MWRA evaluate and gain approval for downsizing the planned secondary sewage treatment plant, which saved Boston area rate payers $160 million.

USGS studies in Boston Harbor have documented that the concentrations of most toxic heavy metals in surface sediments have decreased to about 50 percent of the levels measured in the late 1970s. The continuing long-term observations provide regulatory agencies and the public with clear evidence that the cleanup program, specifically those efforts to reduce contaminants entering the harbor, is resulting in measurable improvement (Long et al., 1995; USGS, 1998b, 1998c, 1998d, 1998h, 1999).

The USGS-MWRA Massachusetts Bay Project and joint funding agreement is a model for state and federal cooperatives. The USGS studies provided scientific information that contributed directly to the success of a major public works project to improve the environment in Boston Harbor and Massachusetts Bay. The Massachusetts Bay Project also provided the USGS opportunities to develop mapping, monitoring, and modeling capabilities for sediment contaminant studies in other U.S. coastal areas.

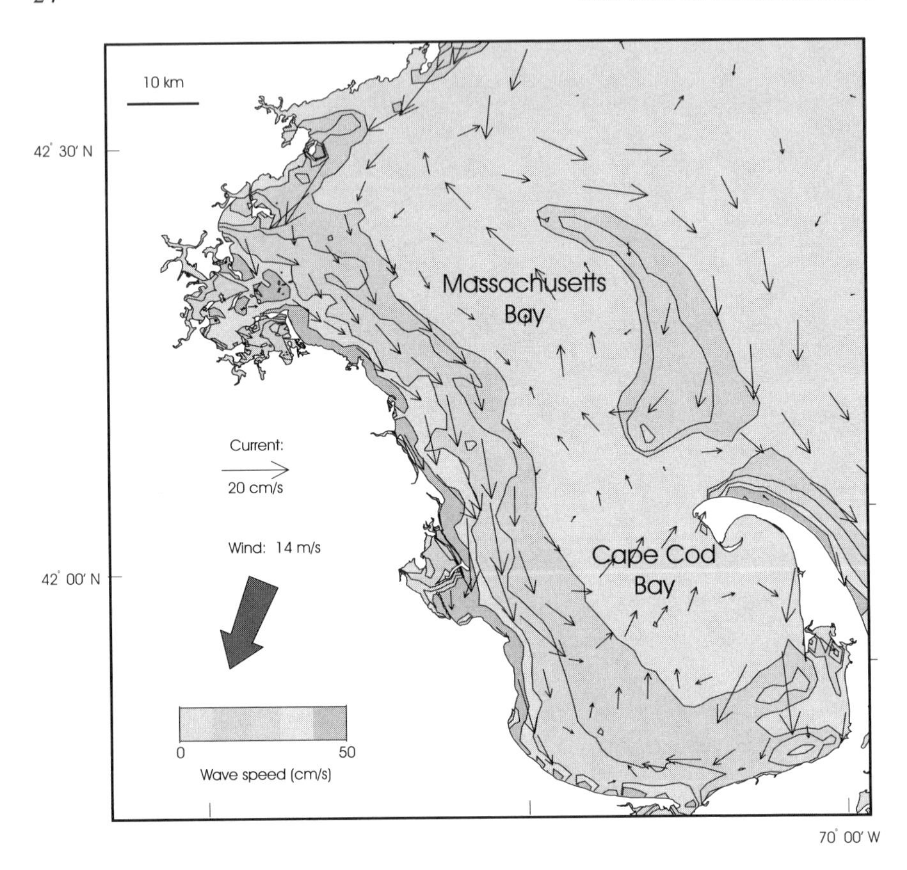

FIGURE 2-2 Modeled wind-induced currents (arrows) and contours of near-bottom wave current speed driven by a northeasterly wind of 14 m/s (28 knots). Near-bottom wave speeds in excess of about 10 cm/s are sufficient to resuspend fine-grained sediments. During major northeasters, fine sediments along the western shore of Massachusetts Bay are resuspended by the wave currents and transported by the wind-driven flow to the southeast toward Cape Cod Bay, where they settle. They are protected from the influence of subsequent storms by water depth and basin geometry. The numerical circulation models provide predictions of the basinwide storm response, which would be very difficult to observe directly (USGS, 1998b).

ridges, chemosynthesis-based exotic biological communities survive at the interface between seawater, groundwater, and hydrothermal vent fluids, respectively. Other environments (e.g., coral reefs) are fragile because they have been impacted in historical times by natural processes and anthropogenic activities. Fundamental studies of the dynamics of estuarine and coastal systems (e.g., Chesapeake Bay), wetlands (e.g., Gulf of Mexico southeast tidal wetlands, San

Francisco Bay, Florida) and regional studies of specific systems that are currently being stressed (e.g., Hawaiian coral reefs) are under way in the CGMP.

Subtheme 3: Marine Reserves and Habitats

In recent years, a number of marine and coastal areas have been designated for conservation and sustainable use. These include national marine sanctuaries, national seashores, national marine parks, areas in the U.S. Environmental Protection Agency (EPA) Bays Program, and other federal coastal and estuarine reserves. Comprehensive geologic scientific information is required to understand the dynamics of these environments and make sound management decisions. In several instances, the management agencies have requested that the USGS provide seafloor bathymetry and sediment distribution maps (e.g., for Stellwagen Bank, Florida Keys, Monterey Bay), which are critical to managing marine reserves. In addition, many biological habitats in the nearshore and offshore regions are being affected by such human activity as waste disposal, seabed disturbance from fishing gear, and overfishing. Understanding the impact of these activities on the seafloor substrate and habitat is critical to managing these economically important areas (Box 2-2), and again the CMGP was asked to apply unique expertise and mapping capabilities.

CMGP research focuses on documenting the changes in biological habitats caused by natural processes and anthropogenic activity (e.g., overfishing on Georges Bank, trawling disturbances on the California coast, snapper habitats off Hawaii).

BOX 2-2
FISHERIES HABITAT MAPPING IN NEW ENGLAND

During the last five years, CMGP seafloor mapping projects in New England's Gulf of Maine have addressed the need to describe biological habitats in terms of their geologic attributes and processes and their importance to fishery species (Butman and Schwab, 1997). Studies have been conducted on Georges Bank and in the Stellwagen Bank off Boston, Massachusetts (Fig. 2-3). The depressed state of the New England fishery required policymakers and managers to develop regulations to conserve and rebuild the fishery by limiting catches and by placing large areas of the seabed off limits to fishing. The 1996 Magnuson-Stevens Fishery Conservation and Management Act requires that the National Marine Fisheries Service and the regional Fishery Management Councils identify and protect essential fish habitats for the most important fishery species. This requirement increased the need for the study of seafloor habitats to conserve fish stocks and to assess and prevent destruction of essential fish habitats by fishing gear (USGS, 1998e).

This issue is addressed through the collaboration of the CMGP, four National Oceanic and Atmospheric Administration offices (National Marine Fisheries Service, National Ocean Service's Coast Survey, National Marine Sanctuaries Division, and National Undersea Research Program), the EPA, the New England Fishery Management Council, and scientists from the University of Rhode Island and the University of Connecticut. The CMGP conducts the mapping and geological investigations and works with biologists in regions where collaborating agencies have identified important environmental and fishery habitat issues.

The National Marine Fisheries Service has a major role in information gathering activities and in the management of the fishery, and it provides financial support for academic biologists and contributes ship time. The National Ocean Service's (NOS) Coast Survey carries out hydrographic mapping and was interested in the utility of multibeam mapping systems for its bathymetric surveys. NOS funded half of the Stellwagen Bank multibeam survey, and National Oceanic and Atmospheric Administration (NOAA) officers participated at sea to train for the new technology, which now has been adopted for its own surveys. The Sanctuaries Division of NOS manages the sanctuary and provides support in the form of ship time for geological and biological sampling in the Stellwagen Bank region. It disseminates results to the public and schools through its outreach program. NOAA's National Undersea Research Program supports research on marine environments and habitats through the use of submersibles and remotely operated vehicles and has supported biological investigations by academic biologists in the mapped regions. The EPA is a joint manager (with the U.S. Army Corps of Engineers) of the Massachusetts Bay disposal site off Boston, Massachusetts, and also is concerned with the past disposal of toxic and radioactive materials in the region. It provided ship time for sampling to aid in interpretation of the multibeam imagery of the disposal site. The USGS role has been to map and characterize the seafloor environments and processes at scales meaningful to biologists and managers. The New England Fishery Management Council makes fishery management decisions and essential fish habitats and identifies important issues that need to be addressed through seafloor mapping and habitat research.

These collaborative studies have identified seafloor processes, species-habitat relationships, and effects of fishing gear on seabed communities that are a basis for regulatory decisions by the New England Fishery Management Council. The council raised the protection level of an important habitat in a presently closed area of Georges Bank by designating it a habitat area of particular concern. The council considered designating a like habitat in another region, but decided that available historic information was inadequate. It is relying on results of a recently initiated CMGP mapping project in Great South Channel to help make a decision. The council now is deliberating how best to reopen a scallop fishery in an area closed to all fishing on Georges Bank. In the face of political pressure to open the scallop fishery, the council has research results to help develop a management plan that will protect the most important groundfish habitats in the area. The council's recent mandate to manage and protect essential fish habitats greatly increases the need for habitat mapping and research.

Environmentalists have become concerned about the effect of human impact, in various forms, on biological habitats. The Gulf of Maine has become the focus of initiatives to establish marine protected areas to conserve biodiversity and rare assemblages. The research efforts described above by USGS and its partners are a major influence on the debate about how to manage marine environments. Finally, the EPA used the results of mapping the Massachusetts Bay disposal site in revising regulations for its management.

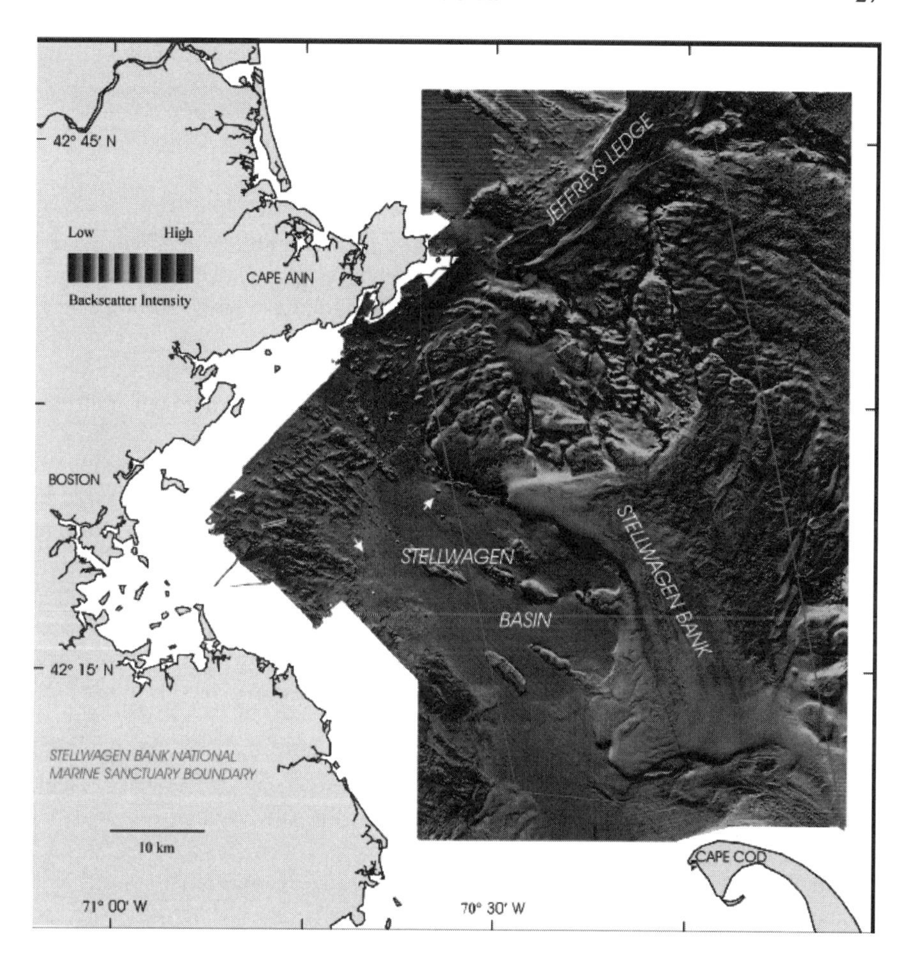

FIGURE 2-3 Sun-illuminated map of Stellwagen Bank National Marine Sanctuary and Massachusetts Bay with backscatter intensity draped over the topography (USGS, 1998e).

Theme 2: Natural Hazards and Public Safety

The overall goal of the natural hazards and public safety theme is to better understand the processes that produce hazards in the coastal and marine environment and their impact on the human population and the natural environment. There is a critical need to better predict the frequency and distribution of catastrophic events that elicit federal response (storms, earthquakes, and landslides); the geologic, human, and environmental consequences of such events; and the local and regional susceptibility to change, such as coastal erosion. Since the types of catastrophic events along different parts of the coastline vary, the studies are necessarily regional in nature.

Subtheme 1: Coastal and Nearshore Erosion

Coastal erosion is a national problem, with enormous economic and social consequences that affect all 30 states bordering the ocean or the Great Lakes. Sediment generation, transport, and redistribution along our coastlines and across the continental shelf are the natural processes by which coastline and nearshore environments evolve. However, the development of large tracts of the coastline as urban and suburban areas, as well as human activities to reshape the coastlines and intercede in the natural riverine, estuarine, and coastal sediment transport processes, have exacerbated the problem of coastal erosion in many areas. The geologic framework of the coastal region and the sediment transport system must be determined in order to understand the problems that must be addressed to maintain U.S. coastlines and to predict the regional effects of any mitigation or management plans. The CMGP is uniquely placed to undertake studies of large-scale geologic processes that shape the coastlines and influence the distribution of sediments along the coast and across the continental shelf (e.g., predictive modeling of bedload transport and cross-shelf sediment transport). In addition, CMGP also addresses local erosion problems and sediment budgets (see Box 2-3), as well as the impact of catastrophic storms and hurricanes (e.g., South Carolina, west central Florida, Louisiana, Hawaii). CMGP has also used new techniques for measuring shoreline position that can provide broad coverage of coastal topography and nearshore bathymetry and can be deployed to key areas before and after major storms (e.g., a global positioning system-based, vehicle-mounted system called SWASH, short for "surveying wide area shorelines," and a coastal and nearshore mapping device that uses a scanning airborne laser, called LIDAR).

Subtheme 2: Earthquakes, Tsunamis, and Landslides

Much of the coastal region of the United States lies in close proximity to the boundaries of major tectonic plates (e.g., the west coast of North America, Puerto Rico and the Virgin Islands, and U.S. territories in the Western Pacific) or is on active volcanic islands associated with hot spots (e.g., Hawaii). These areas are therefore at risk from large earthquakes (both onshore and offshore) and undersea and coastal landslides. These, together with tsunamis that can be generated by earthquake, volcano, or landslide events, pose a serious threat to the growing coastal population centers, and such events continue to cause loss of life and property, as well as disruption to the societal infrastructure. In conjunction with the on-land studies of the USGS Earthquake Hazards Program, the CMGP has a responsibility to advance scientific knowledge of the geologic processes that result in earthquakes and landslides and to provide the scientific basis for deci-

BOX 2-3
EROSION IN SOUTHWEST WASHINGTON

After long periods of shoreline accretion and subsequent private and public development near the coast in southwest Washington state, recent shoreline retreat is putting at risk private property, a public highway, commercial cranberry bogs, community water supplies, wastewater treatment plants, and two state parks (Gelfenbaum et al., 1997; Gelfenbaum, 1998; SWCER, 1997) (Figs. 2-4a and 2-4b).

CMGP is working cooperatively with the State of Washington's Department of Ecology (WDE), with input from local communities on a coastal erosion study funded by the USGS and the state. The USGS co-directs the study, assuring objective and uniform techniques across state boundaries, and conducts some of the research, primarily sediment budget, offshore and geologic framework-related tasks. WDE co-directs the study, interfacing with local communities and other users (state agencies, etc.), performs data management, and maintains the study Geographic Information System (GIS), and conducts some of the research, primarily that related to the beach and shoreline.

Research tasks were assigned to the two agencies based on available technical skills and which tasks would need to continue after the study ended. WDE is being asked to make predictions of future shoreline positions for Washington state parks for two areas using data obtained from the study. The state parks department is using the predictions to plan for the relocation of a campground and to scale back road repairs in a chronic erosion area (Gelfenbaum et al., 1999). WDE is also asked to comment on technical reports dealing with a U.S. Army Corps of Engineers proposed coastal construction plan, an environmental impact statement for a local city's plan to deal with erosion, and a Corps environmental impact statement for proposed dredge disposal. The USGS sat on the governor's Task Force for Coastal Erosion as a representative of the coastal erosion study. The task force recently completed its report, which recommended long-term planning for all coastal communities, conducting an inventory of at-risk infrastructure, and planning for coastal erosion hazards. The USGS and WDE co-produced an educational video on the coastal erosion problem in southwest Washington and on the scientific study that is under way to study the problem. Some 150 copies of the video have been distributed, and a dozen cable TV networks are preparing for showings.

sions on seismic risk, building codes, public disaster plans, and land use and development. To evaluate the potential for large events and to model the likely impacts, CMGP is conducting geological and geophysical work in the Pacific Northwest, the California borderlands, and the northeast Caribbean. It is also analyzing the large volume of data collected from earthquakes to model their hazards. Coastal cliffs are being studied south of San Francisco to determine factors controlling slope instabilities. To assess the risks of disaster and to aid in mitigation planning, regional studies are focused on areas at high risk, providing a geologic framework in the context of historical events and their associated damage.

Columbia River Littoral Cell
Erosion Hot Spots

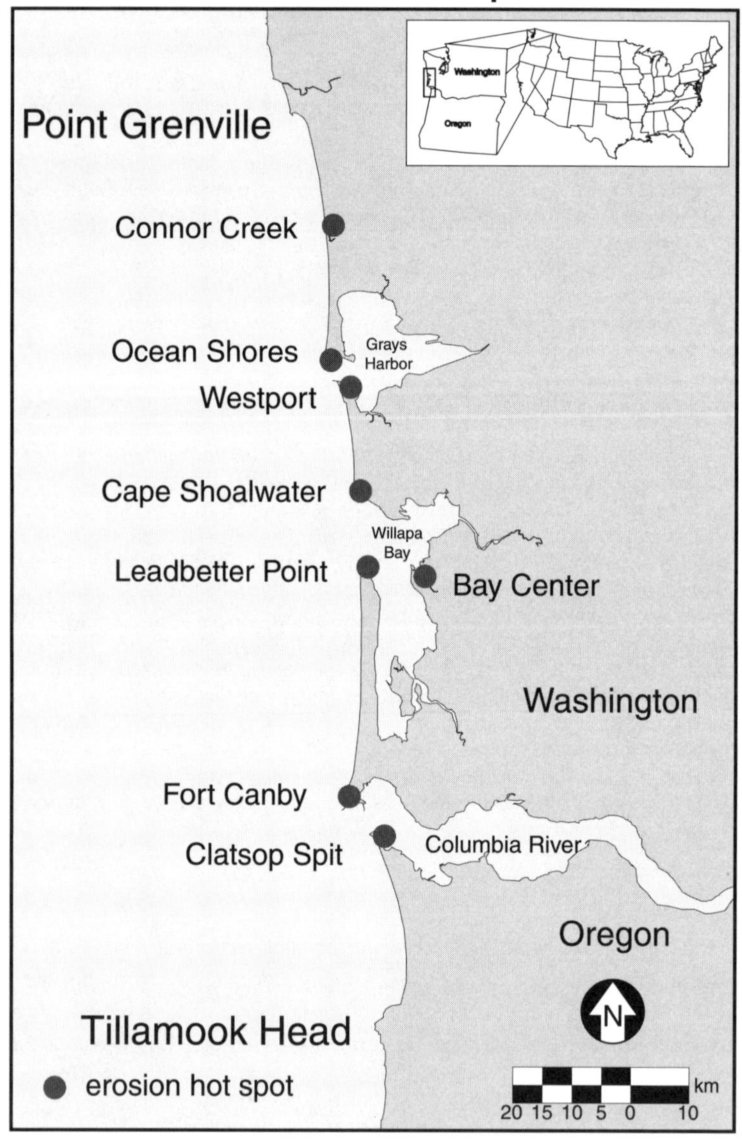

FIGURE 2-4a Erosion "hotspots" in Washington state (SWCER, 1997).

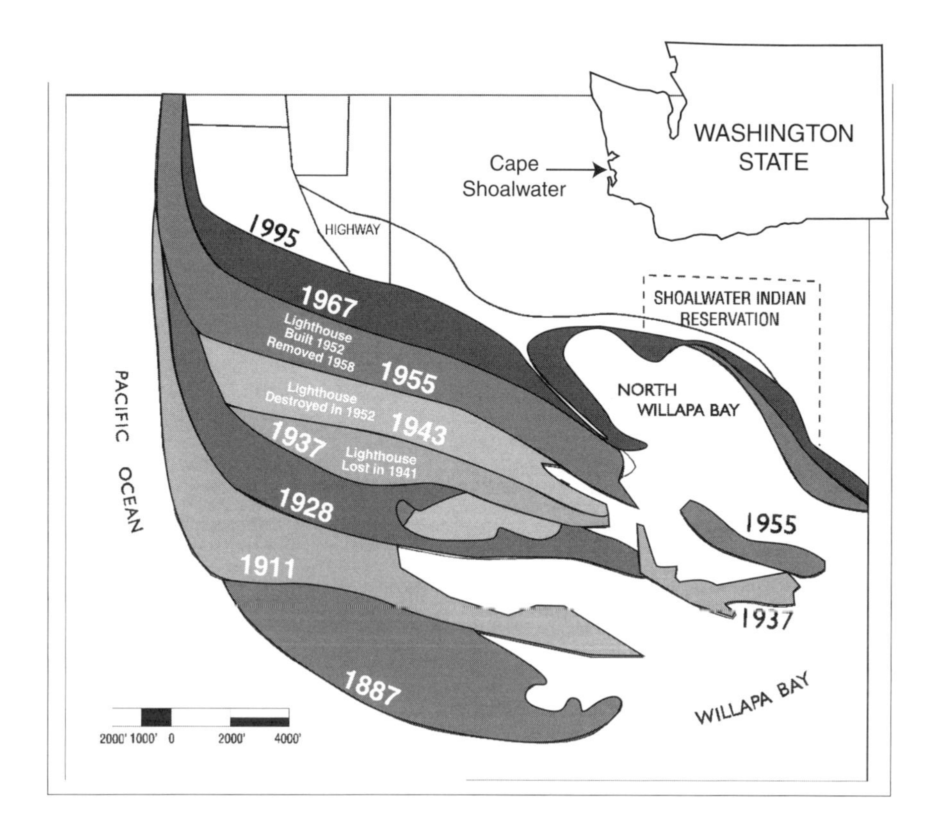

FIGURE 2-4b History of shorline retreat at Willapa Bay, Washington (SWCER, 1997).

Theme 3: Natural Resources

The increase in population in the United States will put increasing strains on the nation's natural resources. The identification of new mineral and energy resources in the marine environment and a better understanding of the geologic controls on subsurface flow of groundwaters in the coastal zone are critical to the management of the supply of resources. CMGP studies of natural resources are broken down into three subthemes (with the objective of understanding the formation, location, and geologic setting of coastal and marine natural resources; the geologic effects of resource extraction and its impact on coastal and marine ecosystems; and how onshore exploration for economic mineral deposits can be refined and broadened, based on information gained from offshore resource occurrences.

Subtheme 1: Water Resources (Coastal Aquifers)

In many coastal communities, declining groundwater levels and increasing demand for freshwater by the burgeoning population have led to intrusion of saltwater into coastal aquifers. It is critical to understand and model the subsurface fluid flow between onshore and offshore aquifers if coastal aquifers are to be managed effectively. Subsurface fluid flow is governed to a large extent by the characteristics of the subsurface geology; hence the geologic framework is required as a basis for determining fluid flow patterns and the chemical reactions between the fluids and the host rocks. CMGP is conducting geologic framework studies of saline encroachment and nutrient transport in coastal aquifers in San Pedro in southern California, the Delmarva penninsula, and in south Florida (Box 2-4).

Subtheme 2: Marine Mineral Resources

This subtheme encompasses a variety of materials that are found in the coastal and marine environment, ranging from sand and gravel needed for aggregate supply and beach replenishment to such metals as manganese, copper, nickel, platinum, cobalt, and zinc, which are critical for industrial and strategic uses. They are the products of dynamic physical, chemical, and biological processes that result in the minerals being concentrated in specific areas. Understanding those processes and their role, the overall geologic system of the coastal ocean, and the EEZ is an appropriate role for CMGP. In conjunction with the land-based Mineral Resources Program, the CMGP also has a regional responsibility to determine the nature and extent of offshore mineral deposits and assess their viability as an economic resource. CMGP focuses on three types of deposits: 1) sand, gravel, and heavy mineral concentrates formed in coastal and shelf areas by sedimentary processes (e.g., Hawaii, Long Island); 2) phosphorites and iron-manganese crusts formed in shelf and deep-sea areas by low-temperature hydrogenic processes (e.g., eastern Pacific, Blake Plateau); and 3) polymetallic sulfides formed in oceanic rift zones, island arcs, and on mid-plate volcanoes (e.g., northeast Pacific ridges, west Pacific island arcs).

Subtheme 3: Energy Resources

Offshore oil and natural gas deposits represent a significant component of the potential energy resources of the United States, so it is appropriate for the CMGP to maintain strong research programs that advance our knowledge of the formation, migration, accumulation, and distribution of oil and gas, particularly as related to the geologic processes that have formed the continental margins (e.g., Gulf of Mexico). In addition, studies in recent years have documented that gas hydrates may be abundant in the sediments on the deep continental margins and may represent an important global reservoir of carbon. The CMGP has a

BOX 2-4
GROUNDWATER IN THE FLORIDA KEYS

Groundwater quality in the Florida Keys is often degraded by processes related to aquifers that introduce freshwater into marine settings or saltwater into freshwater settings or that store freshwater in marine sedimentary reservoirs. In addition, treated sewage is injected into the limestone under the Florida Keys by on-site disposal systems. There are approximately 25,000 septic tank systems, 5,000 cesspools, and 1,000 Class 5 injection wells. Depth of injection wells ranges from 10 to 30 m. Excessive algal growth, coral diseases (Plate 4), and marine grass and sponge mortality are perceived by the local population, NOAA, and EPA to be caused by sewage nutrients leaking from the groundwater on both sides of the Florida Keys. Determining the rate and direction of saline groundwater movement beneath the Keys and Florida Bay is critical to understanding the fate and effects of subsurface waste disposal in the Florida Keys (Shinn et al., in press). CMGP studies conducted there involve the synthesis of regional geologic data; development of models of fluid flow, mixing processes, discharge, and mass and fluid flux; analysis of water chemistry where possible; and development of strategies and technologies for identification of additional potential offshore aquifers (Shinn et al., in press; Lidz, 1997).

As a direct result of CMGP groundwater research, EPA issued a letter to the State of Florida stating that the geology of the Florida Keys is unsuitable for the use of shallow wastewater (Class 5) disposal wells. Two major actions have resulted from this notice:

1. The State of Florida modified regulations for shallow well installation, putting the onus on the landowner to prove that the wells will not contaminate the surrounding marine waters.
2. The County of Monroe (Florida Keys) began searching for funds to install a centralized sewage system. As a result, the EPA is committing several million dollars to install a pilot centralized sewage plant in Marathon, Florida. EPA funding of this plant partially resulted from a white paper prepared by Bill Kruczynski, aided by the Florida Keys Water Quality Technical Advisory Committee, which includes the USGS Geologic Division person spearheading the Florida Keys groundwater research.

Other significant findings resulting from the CMGP study include recognition that there is a rapid exchange of groundwater and surface waters in the Keys that is driven by tidal pumping. In areas where groundwater is saline, injected wastewater is buoyant and rapidly rises to the surface. Furthermore, recent tracer studies have demonstrated rapid migration of Class 5 effluent (sewage) to surface waters (hours to days). These studies demonstrated that tracers were greatly diluted before reaching surface waters and that some phosphorus was stripped from groundwater by the substrate. The long-term ability of phosphorus stripping by the substrate is currently under investigation. Disposal of wastewater from package treatment plants or on-site disposal systems into Class 5 injection wells results in nutrient enrichment of the groundwater that in turn contributes to eutrophication of the surface and marine waters.

significant role in studying gas hydrates through a four-phase research program that includes geophysical surveys to map the distribution of gas hydrates in sediments, laboratory experimentation to define the physical characteristics of gas hydrates, geochemical studies to determine the processes of formation, the composition and stability of gas hydrates, and quantitative well-log evaluation to identify likely gas hydrate horizons.

Theme 4: Information and Technology

A critical function of CMGP is to collect, and make readily available, basic geologic data that can be accessed easily by scientists, policymakers, and the general public. This involves not only using the best-available scientific instrumentation to collect scientific data and maintaining access to ocean platforms on which to work but also synthesizing, managing, and disseminating the information. CMGP is responsible for high-resolution mapping of the EEZ (Box 2-5) and synthesis of the state of knowledge about coastal and marine geologic systems, and **it should be the national source of information about the geology and geologic processes of the coastal and marine environment.**

Subtheme 1: Systematic Mapping of the Coast and Seafloor

Systematic mapping of the coast and seafloor in the EEZ is an important component of CMGP. These activities provide the base maps for many of the studies conducted under the other CMGP themes, as well as for federal, state, and local agencies in their use and management of the coastal and offshore waters. Maps are produced at a range of scales and include morphology, bathymetry, seafloor lithology, and debris due to human activity. Recent advances in processing and imaging of these data and the ability to overlay data sets have resulted in the production of spectacular three-dimensional visualizations of critical regions of the seafloor (e.g., Stellwagen Bank, Lake Tahoe, Hawaii) that have significantly enhanced the geologic interpretation of the seafloor features. Because the coastal and shallow portions of the continental margin areas are most affected by human activities, they will be the major focus of the CMGP in the next decade. This effort has already begun and is focusing on the shelf areas off major urban areas (e.g., New York-New Jersey, Los Angeles). A particular challenge for CMGP is mapping the very shallow (less than 5 m) coastal and estuarine areas that comprise the critical interface between land and sea.

Subtheme 2: Coastal and Marine Information Bank

A priority for CMGP is the maintenance of a comprehensive information bank containing data in forms that are easily accessible and can be used to facilitate management decisions. This requires a plan to deal with a variety of

BOX 2-5
GLORIA—MAPPING THE U.S. EEZ

On March 10, 1983, President Reagan declared an expansion of the sovereign rights of the United States to all natural resources in a zone extending to 200 nautical miles beyond the shoreline. This newly proclaimed Exclusive Economic Zone (EEZ) provided a mandate for further exploration. The general bathymetry was known, but the detailed physiography was not well known. Only with such detailed knowledge of the seafloor could the resource potential and the consequences of exploitation or other activities on the physical, biological, or chemical systems of the seafloor (USGS, 1998f).

In 1984 the USGS launched a program using a long-range sidescan sonar system (Geologic Long-Range Inclined Asdic [GLORIA]) to study the entire EEZ. During the summer of 1984, scientists from the USGS and the Institute of Oceanographic Sciences (IOS) of the United Kingdom surveyed the EEZ off California, Oregon, and Washington, an area of about 850,000 square kilometers. The results of this survey are 36 two-degree sheets, at a scale of 1:500,000. The survey cost approximately one penny per acre. The acoustic images produced by the program are no less remarkable than the first photographs from the far side of the Moon (Chavez, 1986).

A cursory glance at the GLORIA imagery reveals a multitude of geologic features: volcanic edifices, fault scarps, channels, levees, slump scars, large sediment bedforms, crustal lineaments, and textural or tonal differences that reflect varying sediment types. These images provide the framework for a "road map" to direct more detailed investigations. As land surveys commonly rely on various types of remotely sensed data, so the clearer perception of submarine features provided by GLORIA enables marine geologists to focus on specific features of interest (EEZ-SCAN, 1986).

In the late summer and fall of 1985, the USGS conducted surveys of the EEZ in the Gulf of Mexico and around Puerto Rico and the U.S. Virgin Islands. These 1985 surveys abutted an area surveyed in 1982 as part of the outer continental shelf geohazards work that focused on the Texas-Louisiana continental slope and as part of the preliminary work on the Deep Sea Drilling Project in the Mississippi Fan. The collected GLORIA data were processed and digitally combined to produce continuous imagery of the seafloor. The 1982 and 1985 data sets were combined to produce sidescan coverage of the EEZ in the Gulf of Mexico. Sixteen digital mosaics of a two-degree by two-degree (or smaller) area with a 50-meter pixel resolution were completed for the Gulf of Mexico. The mosaics were later combined to produce an overview of the Gulf of Mexico (Paskevich, 1996).

From February to May 1987, five cruises were conducted to cover the Atlantic Continental Margin EEZ seaward of the continental shelf edge, from the Canadian border southward to the northern Blake Plateau off Florida. The innermost Blake Plateau north of latitude 30°N and most of the plateau south of that latitude were not imaged during 1987 because of lack of time. As in earlier EEZ reconnaissance surveys, the USGS used the GLORIA sidescan sonar system to complete the geologic mapping. Twenty-three digital mosaics of a two-degree by two-degree (or smaller) area with a 50-meter pixel resolution were completed for the Atlantic Continental Margin. Twenty-one of the mosaics was later combined to produce an overview of the Atlantic Continental Margin.

material types, including digital data, paper records, reconnaissance images and bottom photographs of the seafloor, and samples. In addition, technical developments are required for digital data rescue and for continually upgrading the archived material to new formats. The goal is to provide the best-possible comprehensive geologic information about the seafloor within the EEZ (Box 2-6). The CMGP in collaboration with other agencies has already developed local databases for specific components (e.g., the database of contaminated sediments for the Gulf of Maine).

BOX 2-6
DINKUM SANDS—APPLICATION OF USGS COASTAL GEOLOGIC KNOWLEDGE TO U.S. SUPREME COURT CASE

From the early 1970s to mid-1980s the CMGP conducted research to characterize the continental margin off Alaska's oil-rich north slope. As part of the these pioneering geologic framework and environmental studies, the USGS conducted research to understand the unique ice-related sedimentary and erosion processes that control and modify the coast, barrier islands, and seafloor morphology in the region, thereby promoting an understanding of environmental factors relating to energy resource exploration and production in the Beaufort Sea.

A by-product of this research was the impetus it provided to the USGS to adjudicate a dispute between the federal government and the State of Alaska over Dinkum Sands, a shoal in the Beaufort Sea about 21.5 kilometers northeast of Prudhoe Bay. The dispute hinged on whether Dinkum Sands was an island, in which case it and 27 oil lease tracts would belong to the State of Alaska, or whether it was an underwater shoal, in which case the area would belong to the federal government. The case was heard by the U.S. Supreme Court, which in 1996 ruled in favor of the United States. CMGP staff input was based on long-term first-hand local and regional knowledge along with credible research studies on ice-related arctic coastal processes that convincingly demonstrated that the feature was indeed an underwater shoal. As a result, more than $1 billion in oil revenue that had been held in escrow and a half million acres of seafloor were awarded to the federal government. The Justice Department singled out the USGS and the lead scientist for recognition and appreciation. At the present time the interest from the Dinkum Sands escrow account is providing over $6 million per year to the University of Alaska's coastal and marine research community. The USGS Biological Research Division in Alaska has responsibility for administering these funds (Grantz et al., 1980).

Subtheme 3: Assessments and Evaluation of the Information Bank

The information bank allows for periodic assessments of the adequacy of knowledge about the coastal and marine geologic environment. Other types of assessments requiring regional syntheses of many data sets to address a specific problem (e.g., contaminated sediments, available resources) are also carried out under this subtheme.

Subtheme 4: Technology and Facilities

Marine and coastal studies require a wide range of scientific instrumentation and access to a variety of platforms capable of operating in both shallow and deep water. Individual investigators in the CMGP develop equipment for specific types of measurements (e.g., sediment transport measurements and current direction and speed), real-time data retrieval, and collection of samples (e.g., the Seabed Observation Sampling System, or SEABOSS). Other equipment, such as seafloor swath mapping systems, is often leased to ensure that the most advanced systems are used to obtain the highest-quality data. The CMGP maintains the capability to process seafloor observations and data and to produce data products rapidly and efficiently. In addition, the CMGP maintains specialized analytical and experimental laboratories (e.g., organic geochemistry lab and the Gas Hydrate and Sediment Test Laboratory Instrument, or GHASTLI) for various types of measurements that are critical to its mission.

THE COMMITTEE'S FINDINGS

The committee believes that the CMGP represents an extremely important component of the USGS Geologic Division. It conducts scientific research and assessments on the dynamic and complex geologic systems that underpin the coastal and marine environments, and **it is the nation's primary resource for geologic information critical to the management of our coastal and marine environments. CMGP is a program that focuses on a region with myriad geologic processes, as opposed to other USGS programs that focus on a single earth process. This regional focus is needed and is not to be found elsewhere.** As such, **CMGP fills a critical niche in the USGS by providing the fundamental geologic studies necessary to describe and manage coastal and marine resources.**

The committee reviewed the recently published document entitled "Geology for a Changing World" (USGS, 1998h), which presents a science strategy for the Geologic Division for the years 2000-2010. The role of coastal and marine geologic studies in the USGS and the inclusion of the coastal areas and EEZ as part of the responsibility of the USGS are not clearly articulated in this document. Unlike other Geologic Division programs that are each explicitly tied to at least one of the division's seven Strategic Science Goals, the CMGP, because of its geographic focus, addresses aspects of all the goals. As a consequence, CMGP does not appear to be an explicit priority for the Geologic Division, as reflected both in the division's Strategic Science Goals and in the entire planning document. Given the multidisciplinary nature of many of the problems facing policymakers responsible for coastal and marine resources and the unique technical and logistical challenges of working in this area, **the committee strongly believes there is a need for a discrete and well-focused program in the**

Geologic Division that is dedicated to dealing with coastal and marine geology. This will provide a venue for interdisciplinary studies of the complex marine and coastal systems that would be difficult to undertake in the division's other topical programs. Such studies require different tools and strategic approaches than those used in land-based studies, and CMGP is scientifically and technically best equipped to conduct them. The committee recommends that the role of the CMGP and its unique niche in the USGS be made more visible in USGS planning publications. **The USGS, at the highest levels, needs to emphasize the economic and societal importance of understanding both the fundamental nature of the geologic framework of the nation's coastal and marine areas and the role of geologic processes in controlling the quantity, transport, and distribution of living and nonliving resources.**

The committee also reviewed the planning documents and current projects of CMGP to determine how clearly the CMGP, its goals, and responsibilities are presented in its publications and in its activities. As presently configured, the CMGP is fragmented into unrelated projects. Although these projects fit into the Geologic Division's Strategic Science Goals on an individual basis, as a group, they do not convey the sense of a coherent scientific effort focused primarily on the geologic framework of coastal and marine areas. The committee determined that there is a need for a CMGP strategic planning process aimed at more strongly identifying the CMGP as playing a leadership role in developing an understanding of coastal and marine geologic processes and providing the geologic framework for science-based management of nearshore and offshore environments. **The committee, therefore, recommends that, as part of the strategic planning process, the CMGP develop a new mission statement that identifies the role of the CMGP in the USGS and clearly articulates its responsibilities.**

The committee also conducted a review of the themes of CMGP to evaluate how well they map to the Geologic Division's seven Strategic Science Goals; the results are presented in Appendix D. Studies within each of the themes are broken down into fundamental studies that are designed to improve understanding of the complex geologic processes in the marine and coastal environments and more regional studies that fall within well-defined subthemes. The overall conclusion of the committee was that individual projects currently being conducted by CMGP map well into the Geologic Division's science goals, but they do not group into coherent scientific efforts in the themes and subthemes. For example, topics in CMGP Theme 1 (Environmental Quality and Preservation) are the sedimentary record of long-term geologic change, the dynamics of natural sediment transport processes, the transport of pollutants, and marine reserves.

The Geologic Division approaches geological studies of the environment from the "hazard, impact, or change" (whether natural or anthropogenic) perspective, which fits well with CMGP Theme 2 (Natural Hazards and Public Safety) and its subthemes. However, in the Geologic Division there are four other programs that address hazards of various types (e.g., the earthquake, volcano, landslides, and global seismic network programs) and how the CMGP efforts dovetail into the larger efforts of these other programs is a concern. Similarly, Theme 3 (Natural Resources) directly addresses the Geologic Division's Goal 3, plus it also incorporates a subtheme of water resources that is not reflected in the general description of Theme 3. This raises the issue of how CMGP studies integrate with the Water Resources Division of the USGS and the Minerals Resources and Energy Resources programs in the Geologic Division. Finally, Theme 4 (Information and Technology) focuses on infrastructure issues that directly address the Geologic Division's operational, rather than scientific, objectives.

Clearly, information management, maintenance of scientific instrumentation, and access to platforms are central to the accomplishment of coastal and marine research and therefore to the successful dissemination of information to the public and to policymakers. These functions are extremely important to the success of CMGP and must be maintained as a critical operational component of the CMGP. However, the committee felt that the definition of these functions as one of the major scientific themes does not fit well with the Geologic Division's Strategic Science Goals. **The committee therefore recommends that the themes and subthemes of the CMGP address the geologic framework component of scientific issues in coastal and marine regions.** While maintaining a close link to the Geologic Division's Strategic Science Goals, CMGP should redefine the themes to address fundamental scientific issues or what the committee concluded are *grand challenges* related to the coastal and marine realm through coherent project groupings. The responsibility for information management, dissemination, and maintenance of scientific instrumentation and platforms does not represent a theme but should be emphasized as a separate but critical function of CMGP. Greater discussion of the possible nature of these grand challenges is the subject of Chapter 3.

3

Future Program Emphasis

The air-sea-land interface that comprises the nation's coastal and marine environments is one of the most important and complex environments on the earth's surface. In this zone, terrestrial, marine, crustal, and atmospheric processes and their interactions operate at various magnitudes and on highly variable time and space scales. Here oceanic and thick continental crusts meet in complex interactions that result in regionally variable uplift and subsidence, changing erosion and sedimentation patterns, volcanoes, and earthquakes. This complex environment forms the coastal areas that provide homes and recreational areas for millions of Americans, as well as the resource-rich continental shelves, slopes and plains of the U.S. Exclusive Economic Zone (EEZ). U.S. continental margins are in a constant state of natural and anthropogenic change and are increasingly being stressed. Earthquakes, coastal landslides, and erosion threaten large population areas along the western margin. Coastal erosion and degradation of biologically rich estuaries are increasing along the eastern continental margin. Rising sea level and rapid subsidence, along with human activities, are destroying one of the largest wetland regions along the northern Gulf of Mexico. In southern Florida and along the tropical islands of the Pacific and Caribbean, the abundantly diverse and rich coral reefs are degrading at an increasing rate. The U.S. continental margins are rich in a variety of living and nonliving resources. It is critical for the United States to have a national program of investigation of the geologic processes that influence these valuable assets. **The study of such complex regions must be framed in terms of the geologic setting and be approached from a systems-science perspective (broad interdisciplinary and**

integrated studies), rather than as a single discipline. The committee believes that the U.S. Geological Survey (USGS) is the appropriate federal agency to lead this effort through integrated efforts of its four divisions.

The committee has identified three *grand challenges* that it believes should form the integrating principle common to all Coastal and Marine Geology Program (CMGP) efforts to fulfill the need for geological information about the nation's coastal and marine environments over the next few decades:

1. establish the geologic framework of the U.S. coastal and marine regions,
2. develop a national knowledge bank on the geologic framework of the country's coastal and marine regions, and
3. develop a predictive capability based on an understanding of the geologic framework of U.S. coastal and marine regions.

To adequately respond to these grand challenges the CMGP must change its structures and procedures. The three grand challenges (discussed in detail below) are intended to provide the CMGP with a long-term focus and are not site or issue specific. Again, these challenges are intended as an integrative principle that should be used to evaluate the relevance of a variety of projects over the next 10 to 15 years (or longer). The resulting investigative program will be varied; as the complexity of the continental margins varies spatially, the underlying need for information will vary temporally, and successful execution of a national investigative program will require a *systems-science* approach. Addressing these challenges will require that CMGP projects make greater use of expertise in other units of the USGS, other federal agencies, and academic institutions. Such expanded interactions should enable CMGP to better communicate the results of its efforts to its user community. Although the committee understands that the variability and complexity of the continental margins is a familiar concept to geoscientists in general, the following discussion is included here to help establish a framework for discussing CMGP's grand challenges and near-term focus areas. It is from this perspective that the committee then argues the value of the grand challenges that face the nation's coastal and marine regions.

THE GEOLOGIC STRUCTURE OF THE CONTINENTAL MARGINS OF THE UNITED STATES

When viewed collectively, the coastal and marine zones of the United States occupy some of the most geologically complex terrain in the world (Plate 5). These areas encompass a wide variety of geologic structures that represent almost the entire range of boundaries identified within the framework of plate tectonics—from mid-ocean ridges off the coasts of Oregon and Washington to subduction zones off the coast of Puerto Rico. This diversity results in differences in the types of geologic processes that have operated, and are still operating, in the nearshore and offshore along the continental margins. Through time, this diversity also results in variations in the distributions of common coastal features, rivers, aquifers, marine and coastal habitat, and marine resources.

This variability is perhaps most easily seen by comparing the geologic structure of the areas making up the continental margins of the United States, which can be categorized into eight major provinces:

Province 1—The Pacific Northwest

A volcanically and tectonically active province that includes a spreading center and a subductive compressive margin. The province is characterized by:

- volcanic and earthquake processes and massive active margin faulting;
- extensive hydrothermal activity along the spreading center, resulting in the formation of metal-rich sulfide deposits and chemosynthesis-based biological communities;
- areas of simultaneous rapid uplift and subsidence;
- a glacially shaped margin with a major river depositing large quantities of sediment that builds the edge of continental margin; and
- strong littoral currents, high persistent wave energy, and periodic tsunamis.

Province 1 is one of the most complex, dynamic, and least understood of the U.S. continental margin. A systems-science view of the area starts with understanding the active oceanic rift generating new oceanic crust. This young, thin crust is being subducted under the continent and the subsequent melt zone forms a line of live volcanoes from northern California to British Columbia. Thus, a tectonic system is operating from the spreading center in the west to the compressive folds under the shelf and beach and to the active volcanoes.

The present coastline runs at right angles across this tectonic grain and therefore the beaches and shelves have a complex history of uplift and subsidence. The active compressional history has resulted in a complex ocean-bottom bathymetry that is host to benthic life of the deep marine and the marvelous tidal pools of the Oregon coast. The landward extent of this complex tectonic system is represented by faults, earthquakes, mudslides, and volcanoes.

Province 2—Central and Southern California

A shearing margin characterized by:

- areas of extremely rapid uplift and subsidence;
- broad continental borderland with active real-time strike-slip faults associated with massive earthquakes;

- numerous submarine canyons that tap a strong littoral river of sand derived from the Sierra Nevada Mountains; and
- moderate wave energy with episodic storm events and periodic tsunamis.

The borderland area of California is one of the great strike-slip shearing areas of the world. Here the Pacific plate meets the North American plate with grinding and sometimes catastrophic results. The result has been an extremely complex margin of isolated, deep basins juxtaposed with uplifted blocks that are islands of shallow banks.

In the north, the Klamath Mountains run to the edge of the sea. The deep lithospheric and crustal structure of the area still holds many secrets of the underlying foundation that are critical for earthquake prediction.

The shearing motions between the two plates have resulted in narrow uplifting and subsiding beaches swept by strong currents and rivers of sand moving along the beaches, which are swept off into the deep offshore canyons. This formed the great deep-sea fan deposits of the area. The Sierra Nevada Mountains have been rising at a high rate and continue to feed sediment to the coastal zone.

So, here as in the Pacific Northwest, the geologic system extends from the escarpment in the west to the Sierra Nevada Mountains to the east. It is imperative that the system be studied as a whole from land to the sea.

Province 3—Western and Central Gulf Coast

A river-dominated coastal system characterized by:

- a large persistent influx of river-borne sediments and freshwater and related density contrasts;
- high sediment loading, which causes rapid subsidence and diapiric salt intrusions;
- a low wave and tidal energy coastal zone that is periodically inundated by intense hurricanes;
- a broad, gentle continental shelf with complex localized salt withdrawal basins;
- broad wetlands supporting a highly diverse ecosystem; and
- an area of major oil and gas production, both onshore and offshore.

In Cretaceous time this province was dominated by massive carbonate reefs along the margin of a new rift basin. Then, as the Rockies rose, huge volumes of sediment were carried south by river systems. This clastic sediment overwhelmed and killed the reefs, producing a wedge of sediment over 13 kilometers thick. Interbedded with these sediments were layers of salt formed during the early history of the gulf. Thus, as this mass of sediment began to slide southward into the newly formed Gulf of Mexico, great down-to-the basin faults formed in the area parallel to the present shoreline. The resulting reefs, salt layers, and organically rich sediments became structurally deformed and formed natural traps of oil and gas.

Province 4—Florida Platform

A carbonate-dominated stable margin characterized by:

- tectonic stability;
- a vast area of modern carbonate accumulation overlying older carbonate deposits;
- low-lying wetlands characterized by broad marshes and mangrove forests;
- low wave and tide energy, storm and hurricane-influenced; and
- a complex coastal aquifer system.

The Florida platform and the adjacent Bahamian platform have an ancient rift history linked to the formation of the Atlantic Ocean. The early extension and rifting led to a complex crust that includes dike-injected continental and volcanic crust overlain by thick reef deposits. The sediment from the rivers of the western Gulf of Mexico did not reach the area, hence prominent reefs exist there today. The warm Gulf Stream sweeps this stable platform, and tropical environments have developed the beaches and shelves in this unique province. This area experiences hurricanes and strong storms that periodically alter the marine environment.

Province 5—East Coast

A passive continental margin characterized by:

- an ancient rift margin;
- a mesotidal system with strong, persistent littoral currents;
- extensive, persistent beach-barrier estuary complexes;
- strong winter storms and passages of intense hurricanes;
- a continuous coastal sand stream; and
- a northeastern section dominated by remnants of Quaternary glaciation and coastal rebound.

The east coast province in the south overlaps with the Florida platform at the wide Blake Plateau. The province extends from the Appalachian Mountains on the west through the coastline, across the continental shelves, and out to the marine slope. The coastal foundation and the related subsidence mechanisms and fault patterns must be understood as a system in order to understand the beaches and shelves.

The central area receives sediment from moderate-sized rivers eroding the old Appalachian foldbelt. This ancient rift margin is underlain by thick sequences of clastic sediment overlaying reefs. To the north, these sedimentary sequences overlie volcanic sections and the New England volcanic seamounts intersect the coast.

The northern portion of this province possesses a coast and shelf that have been modified by the passage of great Pleistocene ice sheets. Since the retreat of these great sheets, the entire region has been slowly uplifted, rebounding from the removal of the great weight of ice. Like the southern portion, this region experiences severe storms, which, when coupled with the regional uplift, create great sea cliffs and a rocky shoreline. The fisheries that occur from the coastal estuaries seaward to the marine banks boast prolific marine life.

Province 6—The Great Lakes

A failed Precambrian rift system characterized by:

- extremely stable tectonics, and
- glacially-dominated landscape.

The geologic setting of the Great Lakes is extremely complex. Lake Superior overlies an ancient failed rift system older than any mountain chain in North America, while Lake Michigan lies adjacent to a great Paleozoic basin and the other lakes over a variety of Precambrian crystalline rocks. Modified by much younger glacial activity, this area's geologic foundation offers the key to understanding the region's natural history and resource potential.

Province 7—Alaska

Our nation's most diverse coastal and marine province is characterized by:

- a southern margin with extreme vertical tectonics, no large rivers, and strong long-shore drift; an ice-scoured northern margin that is an ancient passive margin with drastic seasonal variations in depositional environment; and a western margin dominated by arc-related and strike-slip Tertiary basins that receive huge seasonal influxes of sediment;
- the highest rates of North American vertical tectonics of the Alaskan hinterland arc;
- wetlands that are greatly different from the other provinces because of their dominance by seasonal permafrost and arctic processes; and
- the most diverse and intense natural hazards in the country.

One of the most complex provinces the Alaska province can be subdivided into three distinct geologic realms:

- The southern realm and the Aleutian Islands are part of the extensive Northern Pacific subduction zone. This compressional area is marked by deep trenches, high mountains, and major earthquakes. The area is sculpted by ice and fast-flowing seasonal rivers. Fjords and glaciers are common, and they interact with the environment to form complex ecosystems.

- The Bering Sea marine realm is floored by many subsiding sedimentary basins and hosts the Yukon River delta. The sea ice and runoff features have a strong influence on the marine geology.
- The north coast of Alaska is a geologic system that extends from the Brooks Range seaward to the rifted continental margin. A persistent basement high supports the coastline. Like its southern counterpart in the Gulf of Mexico, organic-rich sediment and geologic structure have created extensive oil and gas deposits across the area.

Province 8—Tropical Island Province

Volcanic islands characterized by:

- highly variable tectonic and volcanic activity in which the dominant sedimentary deposits are biogenic or volcanogenic;
- variable but narrow continental margins cut by numerous submarine canyons bordered by adjacent deep-sea trenches;
- diverse tectonic uplift and subsidence patterns;
- high wave energy and episodic storm events; and
- variable hazards, including volcanic activity, coastal and submarine landslides, and tsunamis.

The tropical islands cover extremely variable provinces from the volcanic island chain of the Hawaiian Islands and other Pacific islands to the Caribbean. Rapid tectonic movement, earthquakes, faults, and volcanoes form the geologic setting for varied ecosystems ranging from reefs and estuaries to deep marine habitats. The beaches are complex in such dynamic settings. The onshore island geology cannot be separated from the offshore analysis of such areas. These beautiful, dynamic areas have many natural hazards in common, including volcanoes, earthquakes, tsunamis, and mudslides.

The distinctly different geologic characteristics of U.S. coastal and marine environments, as well as the variations in oceanographic circulation and weather patterns encountered, result from different geologic processes with diverse spatial and temporal scales that shape the coastlines and seafloor. Hence, understanding the dynamic interface between land, sea, and air and assessing how changes in the coastal ocean might impact ecosystems and human populations requires the determination of the interplay between the fundamental geologic framework of these regions and the more localized natural geologic processes. Furthermore, similarities in geologic processes among the regions help point out areas where understanding developed in one region can be used to advance understanding in another. The CMGP is uniquely qualified to conduct nearshore and offshore marine geologic studies and to integrate the results to produce a national assessment of the geologic structure of the coastal and marine regions of the United States. Specifically, **the committee recommends that CMGP undertake a series of eight regional assessments (i.e., conducted in the eight regions discussed above). These assessments should be designed and conducted in a systematic manner that focuses on differences and similarities among the regions (e.g., the assessments should address geologic processes that operate across region boundaries, as well as those specific to a region). These eight regional assessments should then form the input needed to frame a national assessment.** Such an assessment is the focus of the three grand challenges envisioned by the committee.

Grand Challenge 1: *Establish the Geologic Framework of the U.S. Coastal and Marine Regions*

The CMGP has already compiled excellent regional and local studies but has yet to integrate this information into a comprehensive national assessment of the characteristics of U.S. continental margins. The committee recognizes that this is a change from the current mode, but it will bring a much needed rationale and focus to CMGP research. In addition, this approach will require a rethinking not only of headquarters leadership but also individual scientists at the local centers. A national assessment has to be based on sound fundamental, integrated science (which has been a characteristic of CMPG) but with a broad perspective framed by an understanding of the different geologic settings of the eight provinces of the continental margins. **The committee believes that the CMGP is perfectly poised to answer this grand challenge and therefore recommends that it immediately begin planning for a long-term, integrated, and comprehensive assessment of the nation's coastal and marine regions.**

Although the grand challenge offered here is a thorough assessment of coastal and marine environments, the committee feels that there are several thematic research issues that the CMGP should address as part of the plan to develop a thorough understanding of the entire region:

- tectonic and volcanic processes associated with earthquakes, landslides, tsunamis, and the distribution of mineral resources;
- nearshore and coastal processes associated with shoreline change, biological zonations and habitat changes, groundwater and seawater interface and interaction, transport of contaminated sediments, and the distribution of mineral resources; and
- biogeochemical interactions affecting the mobility of pollutants and the distribution and quality of mineral and energy resources.

The relative importance of these thematic research issues will vary between and even in the major coastal and marine provinces discussed above, but all will require multidisciplinary approaches and alliances with other federal and state agencies. Although efforts to address the three grand challenges will, by their nature, need to be coordinated by program leadership, there should remain some room for individual scientific inquiry.

In setting this first grand challenge, together with its component subthemes, the committee feels that, with recognition of the diversity of the eight major coastal and marine provinces, a greater need for interdisciplinary and national assessments will follow. It is no longer sufficient, for example, to document the erosion rate at a stretch of shoreline in response to storms without taking into account the influence of the associated geologic factors (tectonics, glacio-eustatic rebound, hydrology), which may have a stronger influence on erosion rates. By undertaking a systematic assessment of the geologic framework of coastal and marine environments these interrelated variables can be compared and fundamental causative factors determined. Erosion rates must be integrated with process models, geologic information, and ecosystem models if CMGP is to produce forecasts that lead to sound decisionmaking.

Similarly, wetland loss along the Gulf Coast cannot be examined adequately through single-discipline studies; quantitative information on geologic, biologic, and geochemical settings must be integrated into the studies to make the sound predictions needed to support coastal management decisions. Along the western coastal margin, such coastal changes as landslides and erosion must be integrated into a broad-scale model that takes into account the dynamic tectonic nature of this coast. Lastly, to thoroughly understand and predict the rapid degradation of coastal coral reefs, we must not only study coral ecology but also include studies of the hydrologic, atmospheric, and geologic processes in these environments.

Grand Challenge 2: *Develop a National Knowledge Bank on the Geologic Framework of the Country's Coastal and Marine Regions*

This coastal and marine geological knowledge bank should serve as a comprehensive inventory of geologic data developed by all interested agencies, academic institutions, and state agencies much like the knowledge bank of U.S. oil and gas resources, which has been developed by the USGS energy resource program and the Minerals Management Service. Furthermore, the development of such an inventory would represent a unique opportunity to foster even greater cooperation with federal, state, and local partners.

USGS and CMGP have unique access to many forms of data collected using public funds. CMGP can thus play an important role in making those data publicly accessible. In recent years, this has become somewhat easier to accomplish with the advent of electronic distribution systems (Internet or CD-ROM), but much effort remains to bring some forms of data to the public. In the early years of the twenty-first century, distributed information issues will become more important, and USGS needs to seriously consider its role as an information distributor.

The committee, therefore, envisions a knowledge bank that is far more comprehensive than a simple database or series of World Wide Web sites. The knowledge bank should be developed in Geographic Information System (GIS) format with multiple stacked and interrelated layers of data. Data should be systematically collected at the province scale but would be integrated at the national level. The challenge facing CMGP will be to define the types of layers and then translate them into information and then into a comprehensive knowledge bank. The national knowledge bank should be managed and maintained centrally. Its structure must be designed to support resource management and other science-based decisions by federal, state, and local agencies. Furthermore, this knowledge bank should be designed to become the foundation for the assessment of the health and well-being of the coastal and marine environment.

Building such a knowledge base for wise custodial decisions should begin with the construction of a preliminary data model for each province using all available data and information (at many scales and disciplines). Subsequent gap analysis of data, information, and knowledge would reveal:

- the critical data sets needed to analyze or build a comprehensive data model of the province and
- the fundamental geologic questions that will define the most critical projects and data gathering efforts that are needed to build the data model for each province.

This data gathering leads to or facilitates:

- systematic organization of data and information,
- development of pertinent questions about the geologic framework of the province and its active processes,
- selection and prioritization of projects for developing data that are lacking, and
- communication with other federal and state agencies and state geological surveys leading to cooperative ventures.

Finally, development of a method to derive custom products on demand will likely raise questions regarding competition with the private sector—there are some existing businesses that function as resellers of USGS data, sometimes reprocessed for specific purposes, sometimes not. These are thorny issues that are beyond the scope of this study but that will need to be addressed by the USGS as a whole.

Grand Challenge 3: *Develop a Predictive Capability Based on an Understanding of the Geologic Framework of the U.S. Coastal and Marine Regions*

The third grand challenge reflects the importance of planning to the future environmental and economic health of U.S. coastal areas. Effective planning demands an understanding of the likely scenarios for change to the geologic framework of coastal environments, whether from long-term climate change or from extreme short-term events or human activities.

As pointed out in the recent National Science Foundation planning document entitled *The Future of Marine Geology and Geophysics,* "An important area of future research will be in characterizing and modeling (non-linear) systems in which the input forcing is known or can be measured and the system response can be inferred from the geologic record (geologic time scales) or from direct observation (human time scales)" (NSF, 1999). CGMP, through efforts to address the first two grand challenges, should be in a strong position to lead or contribute efforts to understand the complex and often nonlinear geological processes of coastal and marine environments. **The CMGP should expand and strengthen quantitative model development and change-forecast products to meet management needs for defining the future geologic framework of coastal margins.** This approach is consistent with the pursuit of other grand challenges and with the scientific methods and the principles of adaptive management.

As implied to several times in this and the previous chapter, the committee recognizes that reorganizing CMGP efforts will require that, at least initially, CMGP concentrate its efforts on fewer projects and develop a viable mechanism for identifying the near-term focus and adjusting that focus over time. The following chapter lays out one possible strategy for CMGP.

4

The Federal Role

Through the efforts carried out at each of the regional centers CMGP has developed a rich history of conducting high-quality scientific research on the dynamic and complex geologic systems that comprise the near- and offshore marine environments. Examination of the recent CMGP history suggests that some capabilities have been diminished in recent years by attrition and administrative actions. The committee also found, based on a review of planning documents from the USGS Geologic Division and the CMGP and input from the staff, that no clear and focused identity exists for the program nor does it seem to have a clear mission definition. The committee believes that CMGP, by organizing activities at all three regional centers based on an integrated plan that addresses the grand challenges discussed in Chapter 3, would be well positioned to focus on national, regional, and site-specific coastal and marine issues and problems.

The multidecade time horizon necessary to address the grand challenges means that shorter-term milestones and priorities will need to be established for CMGP. The following sections outline a suggested basis for redefining the CMGP's role and mission in the USGS while addressing the grand challenges.

DEVELOPING NEAR-TERM FOCUS

The recommendations for a long-term CMGP vision in Chapter 3, and the implementation of this vision through development of a strategic plan, will require substantial program refocusing through major changes in proposal development, allocation of finances, nature of personnel, interactions between the centers and headquarters, and many other operational procedures. Such changes will require considerable time to reach the goal of becoming the preeminent federal agency for fundamental geologic information about U.S. continental margins. The committee feels that progress toward these goals can be made in the shorter term by refocusing the scientific efforts ***at all three CMGP centers*** to address a few pressing issues that could serve as the initial steps needed for the grand challenges. Such a refocusing should foster closer interaction of scientific personnel, allow more efficient use of equipment and computer resources, and begin expanding the scale of CMGP research to a national perspective.

Although many pressing issues have been identified, the committee recommends that CMGP concentrate its efforts on understanding the fundamental role of geologic processes in:

- sediment dynamics (erosion, transport, and deposition),
- coastal hazards,
- coastal aquifers and water quality, and
- continental margin habitat mapping and changes.

Shoreline Change and Sediment Dynamics

In 1973, the U.S. Army Corps of Engineers completed an initial assessment of U.S. shorelines. This study is now 26 years old, and more recent studies conducted by coastal states and regional studies compiled by CMGP could form the basis for an updated national shoreline assessment. Many of the ongoing studies at the three centers could be continued and expanded to a more regional scale. These could then be combined with the offshore characterization of the nearshore bottom configurations obtained by high-resolution mapping to more thoroughly tie offshore processes to shoreline change (Figs. 4-1a and 4-1b) and to develop regional sediment budgets. The expansion of this effort at each center would then set the stage for compiling atlases on the health of shorelines and provide a much-needed national database. Eventually, continued study of shoreline change, combined with ongoing fundamental studies of sediment transport, should lead to an assessment of the coastal sediment budget, which is critically important to emplacement of nearshore structures.

Coastal Hazards

One of the more visible and life-threatening aspects facing the high concentration of population along the coastal zones is the wide variety of natural hazards (Plate 6). Volcanic eruptions, earthquakes, coastal landslides, and tsunamis along the western margin and Alaska; subsidence-induced wetland loss, hurricanes, and subaqueous landslides (affecting offshore oil and gas structures and pipelines) along the Gulf Coast; and hurricanes and winter storms along the eastern margin annually cause millions of dollars in damage and tragic levels of injury and death

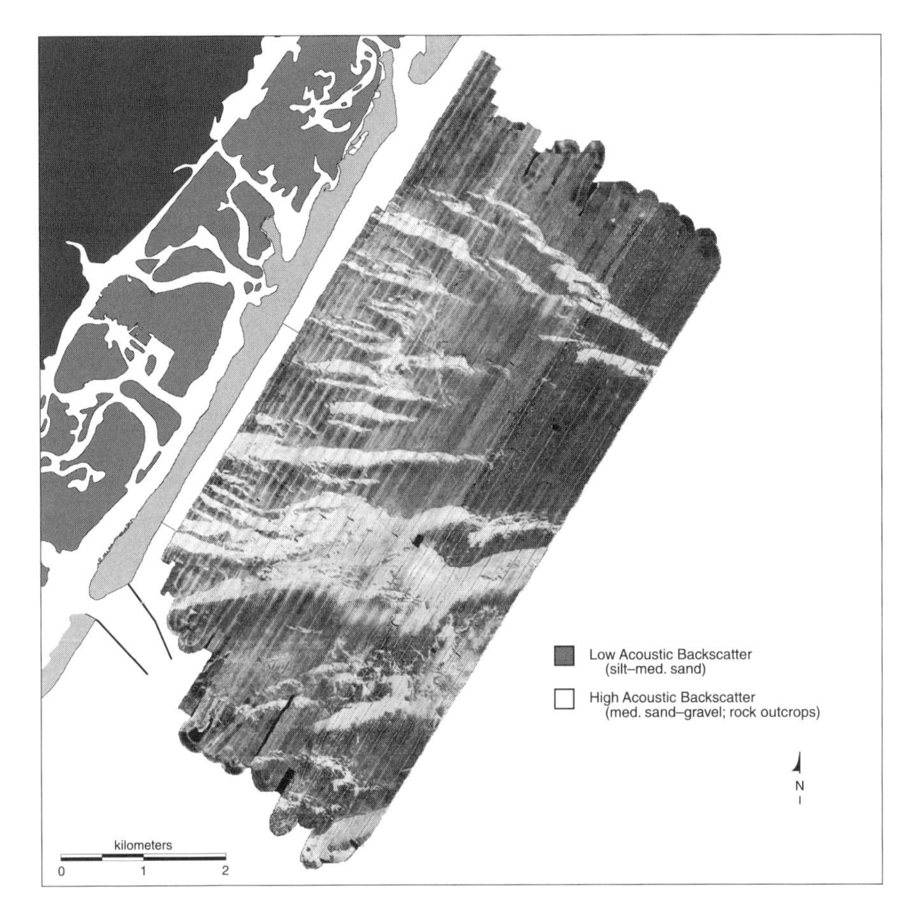

FIGURE 4-1a Sidescan-sonar image of the shoreface and inner shelf off Wrightsville Beach, North Carolina, showing distinct seafloor features that have been used to document sand transport offshore several kilometers from the beach (modified after Thieler et al., 1998).

(Plate 7). Continuation of studies with the National Oceanic and Atmospheric Administration (NOAA) and other federal and state agencies concerning the distribution, magnitude, and timing of coastal and nearshore hazards (Fig. 4-2) could lead to the development of a comprehensive atlas that not only identifies coastal hazards on a national scale but also provides analyses of their magnitudes and recurrence intervals. Creation of this compilation should be thoroughly integrated among the three centers and in the short term should help in developing procedures for standardization of formats that will be so crucial in meeting

FIGURE 4-1b Interpretive geological map of the shoreface and inner shelf off Wrightsville Beach, North Carolina, based on the sidescan-sonar image in Figure 4-1a, as well as seismic data, vibracores, and diver observations (Thieler et al., 1998).

the second grand challenge—development of a national knowledge bank on the geologic framework of coastal and marine regions.

Coastal Water Quality

Two aspects of water quality are important in the coastal zone: (1) saltwater intrusion into groundwater and (2) eutrophication of coastal ecosystems.

Around the perimeter of every continent and island is a coastal zone where continental (meteoric) groundwater meets seawater in the subsurface. Because of the difference in fluid density between freshwater and seawater, the interface between the two fluids extends inland from the coast in the subsurface. Global

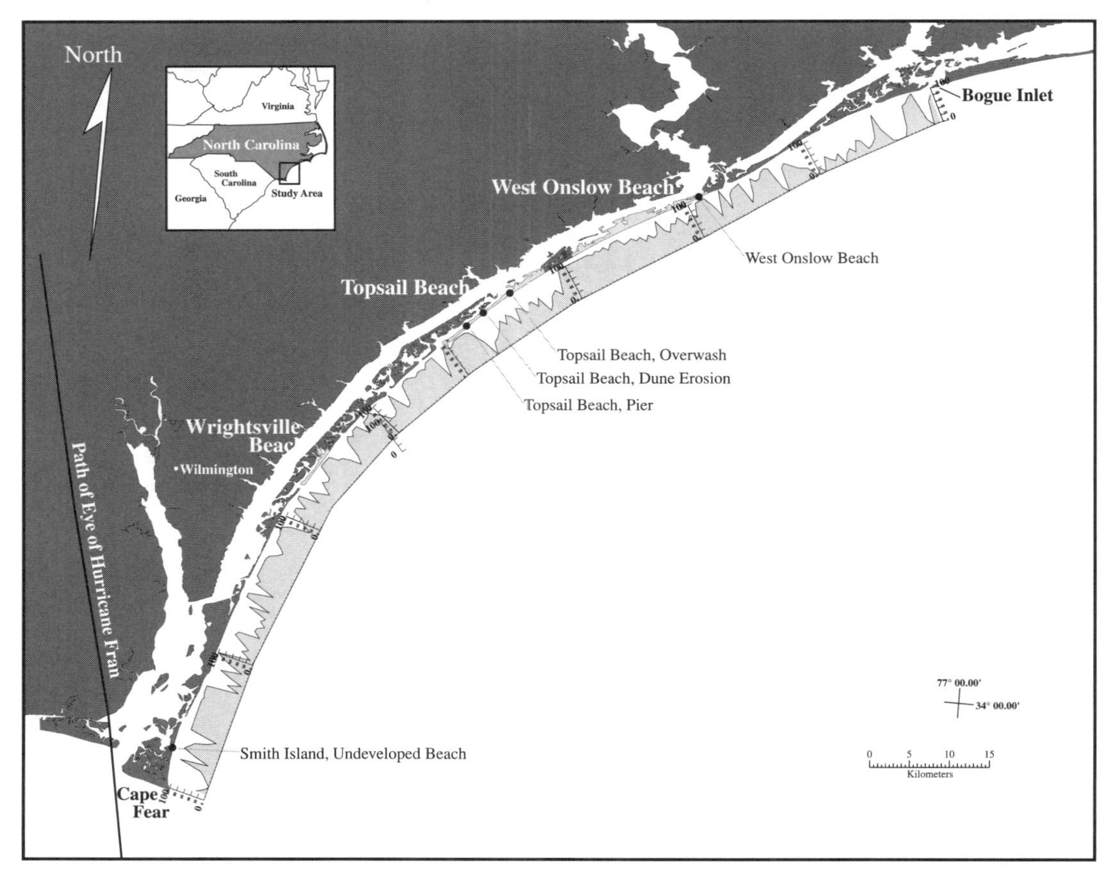

FIGURE 4-2 Preliminary mapping of overwash after Hurricane Fran on September 5, 1996, at Cape Fear to Bogue Inlet in North Carolina (percentage overwash in ½-kilometer increments along the coast) (Stumpf et al., 1996).

warming is most likely to raise sea level 15 cm by the year 2050 and 34 cm by the year 2100. The effect of sea-level rise on the position of the freshwater and saltwater interface depends on the slope of the land in the coastal zone. As coastal land is flooded, the interface in the subsurface migrates inland.

Present production of groundwater and excavation of coastal navigational and drainage canals could possibly have a greater or more immediate effect on seawater intrusion than sea-level rise. Production (extraction) of groundwater that lowers the water table by 1 m, for example, can result in a rise of the interface by 40 m. The intrusion of seawater owing to coastal canals has been well documented in Florida, for example.

Eutrophication of coastal ecosystems is a serious and growing problem in the United States and around the world; for instance, oxygen-poor waters on the inner continental shelf of the northern Gulf of Mexico can extend over an area as great as 18,000 km. Other areas at risk include the Chesapeake Bay; Long Island Sound; San Francisco Bay; portions of the Baltic, North, and Black seas in Europe; and the Harvey-Peel Estuary in Australia. The geographic extent and changing severity of eutrophication, the relative susceptibility of different coastal ecosystems, and the most effective nutrient control strategies are highly uncertain because appropriate monitoring and supporting research are lacking. Uncertainty exists not only in the scientific understanding of eutrophication but also in how to translate existing and future knowledge into good control eutrophication policy.

Effective management must integrate knowledge of the oceanic, estuarine, and watershed processes that contribute to eutrophication. The overall role of coastal geomorphology and the geologic processes that shape it, as well as the role of contaminated coastal groundwater in eutrophication, may be important in understanding why certain estuaries seem to be more susceptible to nutrient pollution and eutrophication. Furthermore, the role of suspended sediment in stressed estuaries must be further evaluated before efforts to reduce nitrogen and phosphorous loading can be expected to yield anticipated benefits.

Continental Margin Habitat Mapping and Changes

The Magnuson-Stevens Fishery Conservation and Management Act as amended by the Sustainable Fisheries Act of 1996 contains essential fish habitat provisions and mandates a supporting research effort to: 1) describe and identify essential fish habitat; 2) identify and evaluate actual and potential adverse effects on essential fish habitat, including fishing-related and non-fishing-related impacts; and 3) develop methods and approaches to conserve and enhance essential fish habitat. Unfortunately, relationships between bottom character and processes and living marine resources have been established only in a rudimentary way for relatively small areas and specific sites. No methods have been developed to apply any of these rudimentary relationships to areas beyond those studied. Finally, there is little understanding of either natural or anthropogenic impacts on the relationship of the seabed to living marine resources.

The NOAA is compiling a national atlas of coastal habitat distribution and change. This undertaking, however, does not reflect the geologic context of the setting that in many instances controls the rate of change. Rates of uplift and subsidence, sediment type, magnitude of processes, and many other variables being compiled by CMGP could indeed enhance this effort. In addition, characterizing the nearshore bottoms that provide the substrate for marine organisms and their habitats would be a natural addition to the NOAA compilation. This integration would allow scientists at all three centers to expand cooperation with another federal agency tasked with providing information about coastal habitats.

Overall

Refocusing the existing program to address a few major issues will require a reallocation of funds and the phasing out of some ongoing research over the next few years. The committee decided not to identify those programs that should be phased out, but it strongly urges CMGP management to concentrate its research efforts and reallocation of funds on a more focused research program, realizing that the near-term focus of the program may change periodically to reflect shifting national and regional priorities. At present, the subset of critical issues identified above seem to be those most appropriate to the overall mission of the CMGP program. **For the next few years these issues should be an integral component of the strategic planning process and should form the basis for a focused research program common to all three centers.** Changing the near-term focus during a long-term effort to address the grand challenges is an important consideration. CMGP will need to select near-term focus projects carefully so as to not become overly focused on near-term issues. One mechanism to avoid over emphasis on issue-driven projects could be to give funding priority to projects that can demonstrate a potential to address near-term issues while providing an understanding of key components of the geologic framework. **In addition, the committee suggests that the concept of fundamental studies be preserved and that funding be maintained at the present level of roughly 10 percent of project funds (Fig. 2-1b). A suite of well-selected fundamental studies should provide CMGP with the flexibility required to address unanticipated changes in national and regional needs for scientific information.** These fundamental studies, therefore, should not be directed at existing issues. Rather, they should represent an opportunity for curiosity-driven research, especially research relevant to the CMGP long-term goals.

ROLE OF CMGP IN THE FEDERAL GOVERNMENT

Although several federal agencies conduct physical science and engineering programs and studies, the CMGP occupies a unique niche by providing capabilities to conduct research and assessments of the geologic processes impacting the nation's coasts. The efforts of the U.S. Army Corps of Engineers are focused on developing engineering solutions to very site-specific coastal problems (e.g., tidal inlet improvement and beach nourishment projects). NOAA's needs for geologic information to address its mission requirements for management of fisheries, sanctuaries, and other coastal resources are not met in NOAA, although the Sea Grant Program does support small geologic research studies conducted by state institutions. The U.S. Federal Emergency Management Agency and the U.S. Environmental Protection Agency rely heavily on the academic community to provide whatever geologic research and knowledge base the agencies require. However, the USGS alone has the ability to frame coastal geologic questions having both regional and national perspectives, while conducting studies that provide the geologic component for interdisciplinary approaches and useful information to decisionmakers. Examples of such efforts include erosion of the southwest Washington coast; Florida Bay information on injection well flushing (Box 2-4); the threat posed by contaminated sediment along the coast of southern California (Box 4-1); and seafloor characterization for essential fish habitat, including coral reef processes.

CONFIRMING THE NICHE FOR CMGP

Collaboration with Federal Agencies

The special expertise of the CMGP in understanding the geology of the coastal oceans lends itself to collaboration with other federal agencies, and some local efforts relating to ecosystems assessments and geologic framework for pollution studies with NOAA, the U.S. Army Corps of Engineers, and the Environmental Protection Agency are being conducted now (e.g., Stellwagen Bank/Boston Outfall, New York Bight dump sites, and Lake Pontchartrain). There is a need for expanded efforts by the USGS to quantitatively describe and model the geologic framework of coastal and marine regions for more effective management of environmental protection and resources. Such collaboration with other federal agencies provides an opportunity for the USGS to leverage its program funds and to amplify the scientific return of investigations carried out by other agencies. **The CMGP should develop a more aggressive approach to collaboration with federal agencies that need information about the geologic framework to meet their mission, including efforts to educate these agencies about the relevance of the information.**

BOX 4-1
PALOS VERDES PROJECT

Historic discharges from the Los Angeles County Sanitation District's ocean sewer outfall contained significant quantities of dichloro-diphenyl-trichloroethane (DDT) and polychlorinated biphenyls (PCBs). Contamination from these discharges is present at high levels in a sewage effluent-affected sediment body on the continental shelf and slope south of the Palos Verdes Peninsula, southern California (Noble et al., 1996; Wong, 1996; Drake, 1996; Lee et al., 1996). The U.S. government and the State of California are suing the parties allegedly responsible for the damages produced by this contamination.

The U.S. Geological Survey (USGS) was asked by the National Oceanic and Atmospheric Administration (NOAA) (lead agency for the Natural Resource Trustees, a group that includes the State of California, U.S. Department of Interior, and other agencies) to conduct a study that would provide information for use in the Palos Verdes lawsuits. Two types of information were requested: (1) maps of the present character and distribution of contaminated sediment on the Palos Verdes margin (Fig. 4-3) and (2) model predictions of how the contamination levels will change in the future. This information complemented other biological, economic, and remediation studies being conducted. To meet this request, the USGS mobilized a major study that included remote sensing, sediment sampling, laboratory analysis of samples, in situ environmental monitoring, and state-of-the-art sediment transport modeling. Results of these studies were reported in expert witness reports and will be presented in a special issue of a research journal.

CMGP staff focused on familiar areas, and outside investigators from universities and industry were brought in to handle other parts of the study. USGS scientists authored the umbrella expert reports, whereas appendixes to the reports were authored by all of the major participants, including USGS and university and industry scientists.

The U.S. Department of Justice is coordinating the overall lawsuit. USGS scientists are presently delivering depositions in the Palos Verdes lawsuit. The USGS data form a major part of this case, one of the largest environmental lawsuits in the country today. Others are also using the data in development of remediation plans.

Collaboration with State Agencies

Existing efforts to collaborate with state agencies and municipalities are important and appear to have been successful. The information provided to the states by the USGS appears to have been used to great advantage to make resource decisions and to establish and update resource management policies (e.g., Boston Harbor Outfall location, Palos Verdes outfall, Florida Bay groundwater, and southwest Washington coast). **It is imperative that the CMGP expand these efforts into regional assessments of the geologic framework of coastal and marine regions. This information is fundamental to the understanding**

FIGURE 4-3 Seafloor map of the continental margin of Los Angeles. The U.S. Geological Survey is using new technology to map U.S. offshore areas. The resulting maps can help researchers and resource managers assess earthquake hazards, study movement of sediment, identify habitats, and understand the offshore environment (USGS, 1998g,h).

and prediction of natural and anthropogenic consequences to the complex coastal and marine environment.

Collaboration with Other USGS Programs and Divisions

The committee finds considerable potential for overlap between the efforts of the CMGP and other programs of the Geologic Division and the USGS (e.g., earthquake hazards, water, energy, and minerals) with concomitant lack of focused efforts in CMGP. There needs to be more substantial partnering with these programs and divisions along the lines of expanded collaboration with federal agencies. The CMGP should focus on the geologic framework questions and assessments while more specific studies are undertaken with its partners. For example, CGMP work on the stratigraphic extent of coastal formations could greatly support efforts by the Water Resources Division to understand aquifer characteristics and evaluate water resources. Along similar lines, CMGP's ability to provide a regional perspective could be essential to the earthquake hazards program in assessing the risk from earthquakes and tsunamis to coastal populations along the Washington coast. **CMGP should make every effort to leverage expertise in other programs and divisions to expand its ability to meet the needs of its diverse user community.**

5

Program Plan Recommendations

As discussed throughout this report, the nation's need for scientific understanding to support sound decisionmaking in the coastal and marine region is real and immediate. The execution of a series of short-term, issue-specific research efforts, such as those currently comprising the Coastal and Marine Geology Program (CMGP), does not offer the greatest potential for developing the scientific understanding needed to allow either the wise stewardship of coastal and marine resources or the ability to anticipate and react to emerging problems and opportunities. Rather, the key to establishing a robust and relevant understanding of coastal and marine areas lies in the pragmatic development of a vision and a set of goals and objectives to guide CMGP efforts to support the Geologic Division and the U.S. Geological Survey (USGS). The committee framed a number of recommendations intended to help the CMGP address many of the points discussed in the previous chapters. The committee restricted its discussion and comments to those aspects of the CMGP organizational structure and procedures (discussed in Chapter 2) that were identified in the statement of task (Box 1-3) or that may impede the CMGP's ability to accomplish its scientific goals.

Although the concept of a vision is easy to understand, its development is another matter. Since the vision should provide guidance for the CMGP for decades, it must be comprehensive, compelling, and attainable. To be successful, however, it should also reflect the mission of the USGS and be relevant to the nation's needs. **A well-crafted vision statement will define goals that are relevant to the actions of every CMGP staff member and to every action undertaken by the CMGP.**

DEVELOPING A STRATEGIC PLAN

Once a clear vision for CMGP has been articulated, the vision must be translated into practical decisions and actions for the CMGP. If the vision represents the long-term goal of the CMGP, then a strategic plan describes how to best achieve that vision. Hence, the successful development and implementation of a strategic plan for the CMGP will provide functional criteria for decisionmaking and further shape the actions and activities that define CMGP on any given time horizon.

The committee strongly recommends that the CMGP undertake the development of a strategic plan. The committee further recommends that the CMGP obtain guidance in the formulation of this plan from its clients and collaborators and from third parties with experience in developing strategic plans. Regardless of the specific mechanisms employed to develop a strategic plan, the committee recommends that the final plan include provisions for:

- attaining the long-term objective(s) embodied by the CMGP vision;
- allowing the periodic development of near-term priorities for CMGP activities in the context of longer-term objectives (similar to the purpose of the current five-year plan);
- implementing mechanisms for short-term (i.e., annual) decisionmaking that will ensure continued and steady progress of the CMGP toward its long-term objective (i.e., help the CMGP achieve its vision); and
- encouraging a continuous evolution of a spectrum of projects and activities that will allow the CMGP to complete activities without stifling the creation of new activities.

VALUE OF STRONG LEADERSHIP

Central to the ability of the CMGP to develop a defining vision of its role, craft a strategic plan for achieving that vision, and refine its approaches and near-term goals will be strong leadership at a number of levels. Strong leadership can provide impetus to the CMGP as it faces a number of challenges in the pursuit of its goals. Effective leadership must reflect the experience, needs, and perspectives of program clients, collaborators, and scientific staff. As with a well-constructed vision, strong leadership will instill a sense of relevance to the actions of every CMGP staff member and every action undertaken by the CMGP.

Conversely, a lack of focused and strong leadership can diffuse the impact of the efforts of the staff and greatly restrict the effectiveness of the overall CMGP. Because leadership is distributed across a number of levels, the failure of the CMGP to achieve its maximum potential cannot be attributed to the actions of any specific individual. Although staff members may be doing their best to accomplish the goals of the CMGP, lack of a central leadership figure will hamper their efforts. **The committee recommends that the USGS, specifically the Geologic Division, implement organizational changes to consolidate and concentrate leadership of CMGP so that it can more vigorously pursue its mission.**

Limitations of the Present Organizational Structure

To focus its efforts, CMGP will have to establish project priorities and change its project mix. The committee questions whether the current organizational structure (Appendix F) will be able to identify and execute the change of focus, as the responsibility and authority for CMGP's performance are vested in a large number of positions. At present, the CMGP coordinator has responsibility for allocating funding but does not have responsibility or authority for staffing and personnel allocations. As a result, effective leadership is difficult to establish and maintain. For example, if a scientific capability needed to address the objectives of the CMGP were identified, there would be no clear and direct way for the CMGP leader to obtain it; instead that person would have to negotiate with science center management.

A more effective model would be the establishment of a direct line of authority for funding and staffing from a program director through team leaders to individual investigators. With a strong director, such a model would provide the necessary leadership and make the CMGP more coherent and relevant to the goals and objectives established in the strategic plan.

Suggested Improvements

The Geologic Division recently instituted a different approach to try to remove budget impediments to accomplishing science. It removed personnel responsibility from CMGP so that the CMGP was not handicapped by having to support scientists who were not contributing to its overall scientific goals. This is an improvement, but the suggestions below on how to organize are very different from the current model.

Advisory Council

Crucial to successful use of the CMGP's limited resources will be timely input and guidance from the CMGP's clients and collaborators. There is no formal mechanism to ensure this input, but the committee understands that the Geologic Division is exploring the possibility of enlarging the size of the CMGP Council (currently comprised exclusively of USGS staff [see Appendix F]) to allow participation of three or four external members (e.g., representatives of state geological surveys, academia, and federal agencies). Although the committee encourages such an inclusion of outside perspectives in this process, it is concerned that, unless the function of the Council evolves to support the CMGP's pursuit of a new vision through the execution of its strategic plan, the value of external perspectives will not be fully realized. **Consequently, the committee recommends that the present CMGP Council be replaced with an Advisory Council charged with new responsibilities and drawn up to reflect the need for broad input to the CMGP.** Specifically, the committee suggests that the advisory council be charged with:

- maintaining CMGP focus and direction through oversight of strategic plan implementation;
- providing advice to the CMGP director (see below) on budget and staff allocations (i.e., identifying near-term priorities); and
- evaluating products and providing feedback to the annual CMGP planning process.

The committee suggests that the Advisory Council advise the new CMGP director and that it be balanced in composition (roughly half USGS and half non-USGS). The Council could include individuals with both technical and policymaking experience and representatives of the regional centers, the Office of the Chief Geologist, the science teams, other USGS divisions, relevant federal agencies and state geological and coastal programs, and academic scientists. To be effective, the Advisory Council should be in a position to review CMGP progress and capabilities, recommend short- and intermediate-term goals, and make broad recommendations to ensure that goals are met.

Program Director

The committee recommends that the Geologic Division structure the CMGP so that leadership (developing, implementing, and ensuring the success of the long-range scientific program) rests with a program director (new position), who is responsible for managing budget and personnel, carrying out the advice provided by the advisory council, and maintaining interagency lines of communication. **This will result in a clear relationship between fiscal responsibility, personnel, and scientific objectives.** Given the broader responsibilities placed on the program director (when compared to those presently vested with a program coordinator), there may be a need for additional senior staff in the program office.

The committee recommends that the CMGP develop and employ an administrative structure with separate lines of authority for the research program and administrative support. This would be similar to the structure typically employed on research cruises whereby the scientific plan is led by the chief scientist and the technical aspects are supported by vessel operations. One approach could be to designate program element or team leaders to manage the day-to-day research activities. In such a structure, the efforts of team scientists could be coordinated and facilitated by the team leader, who, in turn, could answer directly to the program director.

At present, center directors who report to the regional geologist deal with many of the administrative issues faced by CMGP staff. The committee suggests that many of the responsibilities of the regional geologist (see Chapter 1) be vested in the center directors who in turn report to the program director. Such a restructuring would create parallel scientific and administrative lines of authority, which would enhance the CMGP's ability to set scientific goals and then organize the required resources more effectively.

MAINTAINING SCIENTIFIC EXCELLENCE

As indicated in the statement of task (Box 1-3), the USGS is mindful of the valuable role a motivated and experienced scientific staff plays in its ability to fulfill its mission. It is generally recognized that the USGS staff is talented and uniquely positioned to identify major scientific challenges and to design research strategies to address them. Enhanced collaboration with other federal scientists, as well as colleagues in academia and state and local agencies will further enhance CMGP efforts. Unlike their academic colleagues who enjoy great freedom to pursue research in a number of areas, the scientific staff of the USGS recognizes the need to focus USGS resources on research germane to the numerous policy challenges facing the nation. Thus, the ability to maintain a high-quality staff will depend on identifying ways to reward creative and resourceful personnel.

A number of CGMP staff voiced concerns that the current reward system does not adequately recognize efforts that enhance overall CMGP stature but do not result in classic peer-reviewed publications. It is the committee's understanding that much of the emphasis placed on peer-reviewed literature as an indicator of scientific stature is directly related to federal guidelines for evaluating research staff. Specifically, the Geologic Division, like other divisions in the USGS and other federal agencies, has used the Research Grade Evaluation Guide (RGEG) as the primary evaluation tool for basic and applied research positions. As recommended by the Office of Personnel Management (OPM), the division applies the RGEG using a peer panel to assess the research assignment and the researcher's scientific contributions and stature. Through efforts to understand the RGEG, the committee learned that OPM has agreed to review and possibly modify the RGEG. As this issue is not unique to the CGMP, or even the USGS, the committee determined that specific recommendations regarding the RGEG are beyond the scope of this study. **The committee, therefore, encourages individuals at high levels in the USGS to consider the impact the present RGEG may have on its programs and staff and voice those concerns to OPM.** Since the current RGEG applies to a number of federal science programs and agencies, there may be models for its application that reduce potential adverse impacts. **The committee suggests that the GD explore mechanisms (that meet criteria established under the RGEG) to match expectations and rewards across the spectrum of activities and positions in the CMGP (and other GD programs, as appropriate).**

The committee suggests that by focusing the CMGP efforts on a limited number of national and regional efforts, greater cohesiveness can be achieved among the CMGP staff. Furthermore, increased attention to issues of national prominence, and an ongoing commitment to a well-conceived and planned research strategy, should help raise the profile of the CMGP and bring greater recognition to its staff. The proposed benefits of this approach seem to be borne out by the responses received to the staff questionnaire (Appendix C). Long-term commitment to a robust and focused research strategy should encourage staff to make a similar commitment to the CMGP, reducing turnover while encouraging potential CMGP staff members to join the USGS effort. In addition, the important role the scientific staff plays in developing and implementing projects should not be discounted. **Even though many of the recommendations of this report are intended to encourage the development of strong program leadership, that leadership will be most effective if it draws on the knowledge base of its staff.** The advisory council and program director should act as a focusing mechanism for ideas that emerge from the scientific staff, as well as from program clients and collaborators.

Balance Between Full-time and Term Appointments

A review of personnel appointments indicates that in the recent past there has been a large number of term appointments in contrast to full-time hires. Term appointments can be used to acquire special talent required for a particular short-term task or to fill specialized technical staff positions for one-time efforts. However, hiring large numbers of individuals on term appointments can pose a danger to the long-term health of the CMGP. The successful identification of national, regional, and local needs and the generation of projects to address them depend on having a full-time staff that is familiar with both the internal operations of the CMGP and the local and regional agencies that work along the continental margin. It is the contacts made over years of cooperation that usually result in building a solid base of cooperation with state and local agencies. Likewise, it will be full-time appointments that will build a sound coastal and marine program that meets the long-term goal of CMGP.

A review of the age distribution in CMGP reveals that a large number of full-time appointments are nearing retirement age (Appendix E). This is an outstanding opportunity for CMGP administrators to evaluate staffing levels among both scientists and technicians that will be required to pursue the long-term goals in the strategic plan. The new hires should be evaluated in light of the overall

CMGP, rather than simply responding to retirements at the individual centers. In this manner, the program can then respond more readily to the national needs that are so central to the CMGP mission. **The committee recommends that CMGP leadership, during its strategic planning effort, identify the disciplines that will be required to meet long-term goals. Ensuring that these disciplines (that are not represented by collaborators) are well represented during subsequent hiring efforts should be a priority. Furthermore, because these efforts should reflect long-term needs, care should be taken to hire at a consistent and even rate (to the degree possible).**

Budget

The present budget of CMGP is roughly $38 million. Of this amount, approximately $17 million is expended on personnel ($10.1 million) and research operational needs ($7.1 million) (Fig. 2-1b and Appendix E). This is a rather low figure considering that much of the research is field-oriented and that many scientists are involved in more than one project. **It would be to the benefit of the overall program if headquarters staff first evaluated the annual average cost per scientist in other parts of the USGS, other federal agencies, and in academic institutions and compare this to the existing figure in the CMGP. In this manner, the USGS could develop a budget that is in line with the needs of the CMGP.** Such considerations will be particularly important as the CMGP addresses the grand challenges identified earlier. Although a refocusing of existing program funds will be an important step, the pace of progress will be largely determined by infrastructure needs and funds to support key activities.

Technological Advances

Technological advances in computing capability and data acquisition are rapidly accelerating, and maintaining a state-of-the-art equipment pool is becoming a major problem for many agencies. In reviewing the equipment inventory during the visits to the centers, it was apparent that, although the equipment pool was large, many of the large data acquisition systems are quite old and will need to be replaced in the near future. In addition, large amounts of personnel time and funds are being expended on designing and constructing individual systems. With personnel and maintenance costs constantly increasing, it would behoove the various centers to evaluate their long-term equipment needs and to lease rather than purchase equipment. **Efforts should be made to evaluate the overall equipment needs of CMGP programs, based on the proposed strategic plan, and to share equipment among the centers rather than duplicate expensive items.** Leasing from commercial firms, other federal and state agencies, and academic institutions would help relieve increasing maintenance costs, while using new state-of-the-art equipment. There is still a need for some specialized equipment; for example, the equipment designed and constructed in the hydrate laboratory could not be leased or purchased. **It is recommended, however, that decisions to design and build one-of-a-kind equipment be carefully evaluated in light of the overall equipment needs of the CMGP.**

PARTNERSHIPS

Collaborations and partnerships at the program and individual scientist level are critical to the future well-being of the CMGP and its ability to participate in systems-science efforts. Collaboration with other prominent scientists and high-visibility agencies (federal, state, private, and academic) generally leads to high-quality science, allows sharing of resources (personnel, equipment, and ideas), and often results in higher levels of funding. On the individual scientist level, collaboration and partnerships with other scientists promote personal scientific growth and peer recognition.

Collaboration at the Program Level

Scientific advances in solving coastal and marine problems are proceeding rapidly, as a large number of federal and state agencies, private industry, and academia are now focused on this important area. A few decades ago, CMGP scientists, along with a few academic scientists, accounted for a very large segment of the research conducted in the coastal region, and collaboration with the academic community was relatively easy to accomplish. With the increasing number of scientists involved in research along the continental margins (in part resulting from a shift in focus from deep-ocean or blue-water oceanographic research to brown-water coastal research), it is even more important for CMGP scientists to enter into collaborations and interactions with other scientists conducting research in this area. The CMGP derives a great benefit from physical proximity to marine science research centers, major coastal and oceanographic libraries, and industrial technologies. All three centers are located in regions where these capabilities are nearby. It is important for the scientists at all three centers to recognize the inherent benefit of building strong relationships with these entities. Competition in scientific endeavors is healthy, but competing with every agency and academic institution conducting research on continental margins is not beneficial. Instead, **CMGP scientists should make every effort to identify the research efforts of other agencies (federal, state, and local), industry, and academia, to collaboratively obtain or acquire the results, and to integrate them into regional and national assessments.**

It should be noted that the committee distinguishes between seeking out opportunities for collaboration and reimbursable funding opportunities. Although cost sharing could be an important component of collaborative efforts, such collaborations should be clearly relevant to program goals. Reimbursable work that falls outside CMGP goals should be kept to a minimum because it can tie up CMGP assets and key personnel for extended periods of time. It is the committee's understanding that each of the regional geologists often pursues reimbursable work in an effort to help support GD staff that are not fully funded through one or more of the GD programs. It would thus seem logical that reimbursable work that falls outside the CMGP goals would be addressed by non-CMGP staff. Such a distinction could serve both the USGS and potential clients well, as needed work could be completed without diverting the CMGP from its main functions. Furthermore, such a distinction should help minimize potential confusion by avoiding the appearance that two USGS entities are competing for work or otherwise serving similar functions.

Collaborations at the Individual Scientist Level

Collaboration is generally easier to accomplish at the individual scientist level, as individuals with similar interests can often agree on common goals and approaches, and it was apparent that this was being done among the scientists at all three centers. However, this collaboration appeared to be somewhat haphazard and was not coordinated in any manner. Furthermore, because academics can rarely contribute funds, it is generally more difficult for CMGP staff to collaborate with their academic colleagues. However, there are advantages to such collaborations, especially if they make use of expertise or specialized equipment not present in CMGP. **It is recommended that collaboration be stressed at each center and be strongly encouraged by headquarters (i.e., through incentive programs).** Joining with academic partners in writing joint proposals to various funding agencies should be strongly encouraged and, if necessary, rewards should be provided to those personnel who are successful in such undertakings.

ENSURING RELEVANCE TO REGIONAL AND NATIONAL GOALS

Although a large number of agencies and academic institutions are involved in research on continental margins, the CMGP is the only agency with the interdisciplinary scientific resources to characterize the geologic framework of the margins and integrate this information into a comprehensive national assessment. To meet this long-term goal will require a considerable amount of data synthesis by all scientists. Personnel at each of the centers will be required to assemble not only the data acquired by their own research projects but also to use data generated by other federal and state agencies, private industry, and academia. Considerable discussions between scientists and data synthesizers will need to take place to ensure compatible formats and presentation outputs. A review of products created by other agencies (e.g., the National Oceanic and Atmospheric Administration, the National Aeronautics and Space Administration, and the U.S. Navy) needs to be undertaken to determine the most appropriate methods of creating the products needed by federal and state agencies. Once product formats are determined, there is a need to make these products readily and easily available to users. In the opinion of the committee, much of the work presently being conducted by CMGP scientists is of high quality, but the products and data are not easily available and exist in a variety of (sometimes) incompatible formats. Existing products and those that will be generated in the future should be used to educate federal, state, and local governments and the general public regarding the importance of the geologic information. It would be well worthwhile to hire a person at the headquarters level to integrate dissemination activities for the centers and headquarters. Federal budgets are becoming more competitive and without active education about the capabilities of CMGP, as well as active dissemination of its products, the program and eventually the nation will suffer.

References

Butman, B., and W. Schwab. 1997. "Mapping the sea floor off the U.S. east coast." U.S. Department of the Interior. U.S. Geological Survey Information Sheet.

Chavez, P.S. 1986. "Processing techniques for digital sonar images from GLORIA." *Photogrammetric Engineering and Romote Sensing* 57(8):1133-1145.

Cunningham, C., and K. Walker. 1996. "Enhancing public access to the coast through the CZMA." *Current: The Journal of Marine Education* 14(1):8-12.

Drake, D.E. 1996. "Fecal pellets in effluent-affected sediment on the Palos Verdes Margin" (abstract from poster). *EOS Trans. AGU* 76(3). Ocean Sciences Meeting Supplement OS1.

EEZ-SCAN 84 Scientific Staff. 1986. *Atlas of the Exclusive Economic Zone, Western Conterminous United States.* U.S. Geological Survey Miscellaneous Investigations Series I-1792. Scale 1:500,000.

Fluck, P., R.D. Hyndman, and K. Wang. 1997. "Three dimensional model for great earthquakes of the Cascadia subduction zone." *J. Geophys. Res.* 120:20, 539-540, 550.

Folger, D.W., ed. 1996. *Proceedings: 4th Annual Lake Erie coastal erosion study workshop, April 16-18.* Saint Petersburg, Florida: USGS Center for Coastal Geology Open-File Report 96-0507.

Gardner, J.V., M.E. Field, and D.C. Twichell (eds.). 1996. *Geology of the United States' Seafloor: The view from GLORIA.* Cambridge Univ. Press.

Gelfenbaum, G. 1998. "Southwest Washington Erosion Study." U.S. Department of the Interior. U.S. Geological Survey. [Online]. Available: http://coastal.er.usgs.gov/projects98/7242-33780.html [1999, April 19].

Gelfenbaum, G., G.M. Kaminsky, C.R. Sherwood, and C.D. Peterson. 1997. "Southwest Washington erosion study workshop report." U.S. Department of the Interior. U.S. Geological Survey Open-File Report 97-471.

Gelfenbaum, G., C.R. Sherwood, C.D. Peterson, G. Kaminsky, M. Buijsman, D. Twichell, P. Ruggiero, A. Gibbs, and C. Reed. 1999. "The Columbia River littoral cell: a sediment budget overview, coastal sediments 1999." *Proceedings: 4th International Symposium on Coastal Engineering and Science of Coastal Sediment Processes*: 1660-1675. N.C. Kraus and W.G. McDougal (eds.). ASCE, Long Island, NY.

Geological Society of America. 1988. "Gravity anomaly map of North America." Geological Society of America, Boulder, Colo. Scale 1:5,000,000. 4 sheets.

Grantz, A., Barnes, P.W., Dinter, D.A., Lynch, M.B., Reimnitz, E., and E.W. Scott. 1980. "Geological framework, hydrocarbon potential, environmental conditions, and anticipated technology for exploration and development of the Beaufort Shelf north of Alaska, a summary report." U.S. Department of the Interior. U.S. Geological Survey Open-File Report 80-94.

Lee, H.J., D.E. Drake, B.D. Edwards, M.R. Hamer, M.A. Hampton, H. Karl, R.E. Kayen, F.L. Wong, and C.J. Murray. 1996. "Contaminated, effluent-affected sediment on the continental margin near Los Angeles, California" (abstract from talk). *EOS Trans. AGU* 76(3). Ocean Sciences Meeting Supplement OS1.

Lewis, C., and K. Murdock. In press. "The role of government contracts in distributing reinsurance markets for natural disasters." *Journal of Risk and Insurance.*

Lidz, B.H. 1997. "Environmental quality and preservation—fragile coral reefs of the Florida Keys: preserving the largest reef ecosystem in the continental U.S." U.S. Department of the Interior. U.S. Geological Survey Open-File Report 97-453.

Long, E.R., D.D. MacDonald, S.L. Smith, and F.D. Calder. 1995. "Incidence of adverse biological effects within ranges of chemical concentrations in marine and estuarine sediments." *Environmental Management* 19(1):81-97.

Mackey, S.D. 1996. "Relationship between sediment supply, barrier systems, and wetland loss in the western basin of Lake Erie; a conceptual model." *Proceedings: 4th Annual Lake Erie Coastal Erosion Study Workshop, April 16-18.* D.W. Folger (ed.). Saint Petersburg, Fla.: USGS Center for Coastal Geology Open-File Report 96-0507.28-30.

Massachusetts Water Resources Authority (MWRA). 1996. MWRA Technical Report 96-6. 22 pp.

National Oceanic and Atmospheric Administration (NOAA). 1998. *Population: Distribution, Density and Growth* by Thomas J. Culliton. NOAA's State of the Coast Report. Silver Spring, Md.: NOAA. [Online]. Available: http://state-of-coast.noaa/bulletins/html/pop-01/intro.html [1999, April 19].

National Science Foundation (NSF). 1998. "The future of marine geology and geophysics." *Proceedings of a Workshop, December 5-7.* P. Baker and M. McNutt, (eds.). Ashland Hills, Oreg. [Online]. Available: http://www.joss.ucar.edu/joss_psg/project/oce_workshop/fumages/contents.html [1999, June 15].

Noble, M.A., C.K. Harris, P.L. Wiberg, and C. Sherwood. 1996. "Statistical characterizasions of current and wave patterns on the Southern California Shelf off Palos Verdes for use in sediment-transport models" (abstract from poster). *EOS Trans. AGU* 76(3). Ocean Sciences Meeting Supplement OS1.

Paskevich, V. 1996. "Digital mosaics of the GLORIA sidescan sonar data of the Gulf of Mexico." U.S. Department of the Interior. U.S. Geological Survey Open-File Report 96-657.

Reimnitz, E., C.R. Ross, and P.W. Barnes. 1980. "Dinkum Sands." U.S. Geological Survey Open-File Report 80-360.

Sandwell, D.T., and W.H.F. Smith. 1997. "Marine gravity anomaly from GEOSAT and ERS-1 satellite altimetry." *Journal of Geophysical Research* 102(B-5):1039-1054.

Shinn, E.A., C.D. Reich, D.T. Hickey, and A.B. Tihansky. In press. "Determination of groundwater-flow direction and rate beneath Florida Bay, the Florida Keys and Reef Tract." U.S. Department of the Interior. U.S. Geological Survey.

Southwest Washington Coastal Erosion Study (SWCER). 1997. "Scientific study aims to help communities manage coastal erosion problems." February 1997.

Stumpf, R.P., M.D. Krohn, K.L.M. Morgan, R. Peterson, and D. Wiese. 1996. "Preliminary mapping of overwash from Hurricane Fran, September 5, 1996 Cape Fear to Bogue Inlet, North Carolina." U.S. Department of the Interior. U.S. Geological Survey Open-File Report 96-674.

Thieler, E.R., W.C. Schwab, M.A. Allison, J.F. Denny, and W.W. Danforth. 1998. "Sidescan-sonar imagery of the shoreface and inner continental shelf, Wrighsville Beach, North Carolina." U.S. Geological Survey Open-File Report 98-596.

U.S. Bureau of the Census. 1998. *Statistical Abstract of the United States: 1998* (118th edition). Washington, D.C.

U.S. Department of the Interior. 1999. *U.S. Department of the Interior: Protecting the Nation's Coral Reefs*. February 1999.

U.S. Geological Survey (USGS). 1994a. "Selected issues in the USGS Marine and Coastal Geology Program: contaminant transport in Mobile Bay." U.S. Department of the Interior. U.S. Geological Survey. September 1994.

U.S. Geological Survey (USGS). 1994b. "The National Coastal and Marine Geology Program: five year plan for geologic research on environmental, hazard, and resource issues affecting the nation's coastal and marine realms." [Online]. Available: http://marine.usgs.gov/marine-plan/html [1999, June 22].

U.S. Geological Survey (USGS). 1997. "The National Coastal and Marine Geology Program: an updated five year plan for geologic research on environmental, hazard, and resource issues affecting the nation's coastal and marine realms." [Online]. Available: http://marine.usgs.gov/natplan97/natplan.htm [1999, June 22].

U.S. Geological Survey (USGS). 1998a. "Cascadia earthquakes and tsunami hazard studies." [Online]. Available: http://walrus.wr.usgs.gov/docs/projects/cascad.html [1999, April 19].

U.S. Geological Survey (USGS). 1998b. "Predicting the long-term fate of sediments and contaminants in Massachusetts Bay." U.S. Department of the Interior. U.S. Geological Survey Fact Sheet 172-97. February 1998.

U.S. Geological Survey (USGS). 1998c. "Predicting the impact of relocating Boston's sewage outfall—effluent dilution simulations in Massachusetts Bay." U.S. Department of the Interior. U.S. Geological Survey Fact Sheet 185-97. March 1998.

U.S. Geological Survey (USGS). 1998d. "Metal concentrations in sediments of Boston Harbor and Massachusetts Bay document environmental change." U.S. Department of the Interior. U.S. Geological Survey Fact Sheet 150-97. March 1998.

U.S. Geological Survey (USGS). 1998e. "Mapping the sea floor and biological habitats of the Stellwagen Bank National Marine Sanctuary Region." U.S. Department of the Interior. U.S. Geological Survey Fact Sheet 078-98. May 1998.

U.S. Geological Survey (USGS). 1998f. "GLORIA imagery of the U.S. EEZ." [Online]. Available: http://walrus.wr.usgs.gov/gloria/gloria_doc.html#references [1999, April 19].

U.S. Geological Survey (USGS). 1998g. "Perspective view of Los Angeles Margin." [Online]. Available: http://marine.usgs.gov [1999, April 19].

U.S. Geological Survey (USGS). 1998h. "Geology for a changing world." U.S. Department of the Interior. U.S. Geological Survey Circular 1172.

U.S. Geological Survey (USGS). 1999. "Contaminated sediment database development and assessment in Boston Harbor." U.S. Department of the Interior. U.S. Geological Survey Fact Sheet 078-99.

University of Washington Geophysics Program. 1998. "Earthquake history of Cascadia (Washington, Oregon, and Southern British Columbia)." In *Cascadia Database 1793-1929*. [Online]. Available: http://www.geophys.washington.edu/SEIS/PNSN/HIST_CAT/ [1999, April 19].

Wong, F.L. 1996. "Heavy mineralogy of effluent-affected sediment and other deposits, Palos Verdes Shelf, Southern California" (abstract from poster). *EOS Trans. AGU* 76(3). Ocean Sciences Meeting Supplement OS1.

APPENDIXES

Appendix A

Committee and Staff Biographies

COMMITTEE MEMBERS

Joan Oltman-Shay is a senior research scientist for Northwest Research Associates, Inc., and an affiliate of the School of Oceanography, University of Washington. Dr. Oltman-Shay's areas of research include nearshore and inner shelf physical oceanography, such as wind- and wave-driven stresses; infragravity (surface) wave climatology, generation, and dissipation; and mean alongshore currents and current instabilities. She received her Ph.D. in oceanography from Scripps Institution of Oceanography in 1986.

James Coleman received his Ph.D. in geology from Louisiana State University in 1966. Currently a Boyd professor for the Coastal Studies Institute of Louisiana State University and Agricultural and Mechanical College, Dr. Coleman's research interests focus on relationships among process, form, and sedimentary characteristics of recent environments, especially deltaic and offshore regions.

Arthur Green, a research scientist for the Exxon Exploration Company, received his M.S. in geology from the University of Oregon in 1962. His research involves the evolution of the Arctic and its hydrocarbon potential, integrated basin analysis methods, and regional tectonic analysis.

Susan Humphris received a Ph.D. in chemical oceanography in 1977 from the joint program of the Massachusetts Institute of Technology and the Woods

Hole Oceanographic Institution. She currently serves as a senior scientist in the Department of Geology and Geophysics at Woods Hole Oceanographic Institution. Dr. Humphris' research interests include the volcanic and tectonic controls on the distribution and characteristics of hydrothermal activity at mid-ocean ridges, the geochemistry of rock-water interactions, and the role of the associated hydrothermal fluxes in global geochemical mass balances.

Curt Mason, a senior coastal oceanographer with the NOAA National Centers for Coastal Ocean Science, earned his M.S. in physical oceanography from Texas A&M in 1971. His primary research areas are beach erosion, storm impacts, and tidal inlet processes. Other activities include coordinating and prioritizing budget initiatives for new efforts in natural disaster reduction for the National Weather Service, National Ocean Service, National Environmental Satellite Data and Information Service, and Office of Ocean and Atmospheric Research.

Neil Opdyke is a professor in the Department of Geology at the University of Florida. His primary areas of expertise include geology, paleomagnetism, and the evolution of the seafloor. Dr. Opdyke is a member of the National Academy of Sciences. He received his Ph.D in geology in 1958 from Durham University in England.

Nancy Rabalais obtained her Ph.D. in zoology from the University of Texas at Austin in 1983. She is a professor at the Louisiana Universities Marine Consortium (LUMCON). Dr. Rabalais' research interests include biological oceanography, specifically continental shelf ecosystems influenced by large rivers, as well as estuarine and benthic ecology, environmental effects of habitat alterations, and environmental physiology.

Noel Tyler received his Ph.D in geology from Colorado State University in 1981. He is the director of the Bureau of Economic Geology at the University of Texas at Austin. Dr. Tyler's research interests focus on sedimentary geology and oil and gas resource evaluation and recovery optimization in complex reservoir systems worldwide.

STAFF

Dan Walker received his Ph.D. in geology from the University of Tennessee in 1990. He is currently a senior program officer with the Ocean Studies Board of the National Research Council. Dr. Walker's interests focus on the value of environmental information for policymaking.

Shari Maguire received her B.A. from Miami University in 1994. She currently serves as a senior project assistant with the Ocean Studies Board. Ms. Maguire is studying biological sciences at the University of Maryland in preparation for medical school.

Jodi Bachim received her B.S. in zoology from the University of Wisconsin-Madison in 1998. She is currently a project assistant with the Ocean Studies Board.

APPENDIX B

Acronyms and Abbreviations

AGU	American Geophysical Union
BRD	Biological Resources Division
CINDI	Center for Integration of Natural Disaster Information
CMGP	USGS Coastal and Marine Geology Program
DDT	dichloro-diphenyl-trichloroethane
DOI	U.S. Department of the Interior
EEZ	Exclusive Economic Zone
EFH	Essential Fish Habitat
EIS	Environmental Impact Statement
EPA	U.S. Environmental Protection Agency
FEMA	U.S. Federal Emergency Management Agency
FUMAGES	The Future of Marine Geology and Geophysics
FWS	Fish and Wildlife Service
GD	USGS Geologic Division
GEOMAR	Forschungszentrum fuer Marine Geowissenschaften
GHASTLI	Gas Hydrate and Sediment Test Laboratory Instrument
GIS	Geographic Information System
HAPC	Habitat Area of Particular Concern
IOS	Institute of Oceanographic Sciences, United Kingdom
JOIDES	Joint Oceanographic Institute for Deep Earth Sampling
MBARI	Monterey Bay Aquarium Research Institute
MMS	Minerals Management Service
MSFCMA	Magnuson-Stevens Fishery Conservation and Management Act

MWRA	Massachusetts Water Resources Authority
NAS	National Academy of Sciences
NASA	U.S. National Aeronautics and Space Administration
NCGMP	National Cooperative Geological Mapping Program
NGDC	National Geophysical Data Center
NGOs	Nongovernmental Organizations
NMFS	NOAA National Marine Fisheries Service
NMSD	NOAA National Marine Sanctuaries Division
NOAA	National Oceanic and Atmospheric Administration
NOPP	National Oceanographic Partnership Program
NOS	NOAA National Ocean Service
NPS	National Park Service
NRC	National Research Council
NSF	National Science Foundation
NURP	NOAA National Undersea Research Program
NWS	National Weather Service
ODGS	Ohio Division of Geologic Survey
ODNR	Ohio Department of Natural Resources
ODP	Ocean Drilling Program
OMB	Office of Management and Budget
ONR	Office of Naval Research
OSB	Ocean Studies Board
OSDs	On-Site Disposal Systems
PAC	Program Advisory Council
PCBs	polychlorinated biphenyls
PI	Principal Investigator
ROVs	Remotely Operated Vehicles
SEABOSS	SEAbed Observation Sampling System
SATRATAFORM	The Origin of Marine Stratification
SWASH	Surveying Wide Area Shorelines
USACE	U.S. Army Corps of Engineers
USGS	U.S. Geological Survey

Appendix C

Selected Responses to USGS Staff Questionnaire and Clients and Collaborators Questionnaire

1) Scientific Problems

A) What are the three most important scientific problems in the coastal and marine environment that you think the USGS should address over the next 5-10 years?

- The science mission of the USGS is integrative and descriptive, implicitly being to provide a long-term comprehensive view of the nation's realms for use by all in the conduct of life, and in wise decisionmaking. What must be judged is the quality and effectiveness of the USGS in bringing a day-by-day understanding of the nation's realms into the common parlance and awareness of all. This is not what has been commonly believed or practiced, but should be. Therefore, the most important scientific problems of the USGS for its usefulness and survival are not discrete scientific problems but problems involving scientific integration and comprehensiveness over time.
- A lack of a comprehensive long-term, systematic approach to understanding the coastal and marine realms.
- A lack of scientific prediction, providing alternate scenarios from which the public can select, describing the future of the coastal and marine realms.
- Information for science and management. A wide variety of data sets (geographic, geologic, biologic, physical, and chemical) are needed to address the multidisciplinary issues in the coastal zone. USGS is in a unique position to develop, exercise, and maintain some of these data sets.

- Contamination ("toxification") of the environment, including waste disposal and remediation (distribution, transport, fate, and effects of pollutants). Contamination of the coastal environment is widespread (both in the U.S. and worldwide), especially near major population centers and at sites used for waste disposal. Pressures will increase with a growing coastal population and use of the coastal ocean. USGS has a unique role, since many contaminants introduced to the coastal ocean are associated with particles.
- Biodiversity (including declining productivity, disturbance of habitats, protecting habitat, etc.). Fish stocks are declining due to overfishing, habitat disturbance, and other factors. There is increasing pressure for coastal aquaculture. There is worldwide concern for loss of biodiversity, much of which occurs in the ocean or the coastal ocean. The land and seafloor (topography, microtopography, and sediments) play a key role in the habitat of many species; thus, a description and understanding of how these habitats provide shelter and food for species is critical for management and protection of these resources.
- Sea-level rise. Predictions are for global warming to cause a rise in sea level of tens of centimeters over the next century. This increase will have major effects on many coastal communities worldwide.
- Long-term indicators of environmental change. There is a need to develop and maintain indicators of environmental change over the long-term. Long-term observations must be obtained in the context of the overall system and continually analyzed to ensure quality and to further understand the processes causing change (natural and anthropogenic).
- National scope (i.e., we can address issues around the country in a coordinated way);

    — Systemwide regional focus. USGS is not constrained by local funding and can address the issues at an appropriate system level. This is often particularly important in coastal regions where many issues are local (clean drinking water, waste disposal, etc.), yet these problems are best addressed in a regional context.

    — Long-term. USGS can provide a long-term focus for issues and interpretation. This is particularly important in assessing long-term environmental change and in developing and maintaining information and knowledge.

    — Stewardship of data (developing, maintaining, distributing, using).

    — Unbiased, public domain, basic science and resource assessments within a regional framework. We are uniquely situated to walk the fine line between engineering/consulting, academic, and political "agendas." No other agency can do that in the coastal zone.

- The Big Picture: Integration and synthesis of scientific knowledge about the coastal and marine environment, especially with respect to its interactions with human activity.
  - Communicating the big picture: Successful communication of the big picture in such a way that federal and local government agencies, as well as fishermen and other citizens, can make wise decisions with respect to their use and stewardship of the coastal and marine environment.
  - Identifying holes in the big picture: Once the integration and synthesis are under way, some glaring deficiencies in our scientific knowledge will become evident, which, combined with our society's values, will allow rational prioritization of future research and monitoring (much of which should be carried out by universities and private firms).
- Build a knowledge bank for the coastal ocean. The regional studies carried by the CMGP collect multidisciplinary data (geology, oceanography, geophysics, chemistry) and information, often on a regional scale. These large-scale databases and interpretations are applicable to many scientific issues, and as regional studies are completed, the overall coverage and view of the coastal ocean grow.
- Marine hydrology. We know surprisingly little about the hydrology of the continental shelves. Significant amounts of freshwater are likely to be discharging off both coasts, yet we cannot identify where these discharges are. As our aquifers are being altered in so many ways, it is almost inevitable that we will need to understand their offshore extensions.
- The role of coastal and submarine groundwater discharge in delivery of nutrients, toxins, and other dissolved constituents to the ocean.
- Ground and surface water problems: contributing to solutions of problems that are the primary responsibility of WRD and state organizations in the coastal realm and addressing the offshore aspects of the problem.
- Understanding the processes involved in the drawdown of freshwater aquifers due to anthropogenic activity in coastal areas, and understanding their effect on ecosystems.
- Development of better predictive models for shoreline erosion and continental shelf and slope seafloor instability.
- Understand sediment and contaminant transport in the coastal ocean. Develop a regional predictive capability for sediment and contaminant transport and fate in the coastal ocean.
- Coastal erosion and saltwater intrusion of coastal aquifers, defining risk of building and living in coastal zone and mitigation strategies, ensuring that marine environmental policy and regulation involve or are based on credible science.

- Model and predict coastal hazards and their relation to long-term changes in sea level.
- Determine the mineral resource potential of the EEZ of the United States, its possessions, territories, and commonwealths: What is there, where is it, how did it form, and what are the potential resources?
- Link the seafloor environment to habitat. This involves understanding how the seafloor (roughness, texture, 3D structure, chemistry, anthropogenic and natural disturbance, near-bottom flow, animal life history, etc.) makes the seafloor habitat a productive environment.
- Essential fish habitat (EFH) provisions and mandates a supporting research effort. The provisions require a program of research that will provide information to describe and identify EFH, to identify and evaluate actual and potential adverse effects on EFH (including both fishing-related and non-fishing-related impacts), and to develop methods and approaches to conserve and enhance EFH.

I) Map and characterize the entire U.S. EEZ.
— Develop a bottom characterization classification scheme.
— Map and identify resources and extent of impacted area.
— Monitor change due to natural and anthropogenic causes.
— Identify areas of special interest (like national parks).
— Provide information for the conservation of resources.

II) Relate bottom character to biological resources.
— Identify essential fish habitat (EFH).
— Understand physical and biological processes that affect bottom character or the distribution, abundance, and health of living resources.
— Develop indicators of disturbance and recovery.
— Develop models for predictive management that enable application of site-specific information to unknown areas.
— Provide information for the protection, restoration, and maintenance of aquatic coastal and marine habitats, including EFH and their living resources.

III) Assess the effects of coastal development on the habitat of living marine resources.

B) What can be done to facilitate addressing these problems?

- Internal reviews should be used less as a screening and competitive process and more as a guidance and incubator process to see where our limited and declining internal resources should be best used to find new areas of growth. Having outside reviews and non-USGS participants on these panels is crucial.
- Explicitly identify the big picture as our fundamental mission. Identify

research scientists who are capable of carrying out large-scale integrative and synthesizing science, as well as staff members who are skilled at communication, and get them started.

- Federal government agencies should focus more on issue-driven (focus on a problem of societal relevance) rather than curiosity-driven (focus on a natural scientific problem that may or may not have bearing on a societal issue) research. Both issue- and curiosity-driven investigations are forms of knowledge-driven research; doing issue-driven research does not imply that knowledge is not being advanced. There needs to be a balance between issue- and curiosity-driven research in a federal science agency. However, my view is that universities should focus more on curiosity-driven research and federal agencies more on issue-driven.
- Until the reward and promotion system is changed to value products other than papers in peer-viewed professional journals, scientists will continue to focus on and value more highly traditional research. It is my belief that a vital and unique element of our overall responsibilities as federal government research scientists is to provide unbiased scientific information for national policy.

C) Are there any impediments to CMGP addressing these problems?

- CMGP would be able to find a better niche if it established a cooperative program with the states as MMS has done.
- Regionality and bureaucratic programism and divisionism; reluctance to pursue work in politically controversial or administratively complex areas; and unclear boundaries between USGS, NOAA, EPA, USACE, and private contractors.
- Distrust between the director's office, the budget office, the division, and the program. As individual scientists, it is hard for us to address this issue, but as an outside review panel, you have an opportunity to highlight this issue, which I believe has a direct bearing on the performance of our part of the organization at the project level.
- The culture and expectations of the CMGP scientists. We expect to be engaged in the same sort of work as academic scientists or perhaps consulting-firm scientists, with the advantage of a hard money base. Current staff has not been recruited or rewarded for taking the larger view.
- Starts with the reward system. Only curiosity-driven research and the resulting traditional science journal products are valued in this organization.
- We have been directed to publish more in traditional science journals (which the outside end-user would not typically read or use).
- There is no incentive to do issue-driven or relevant research.

- Mindset of current leadership and limitation of funds.
- A lack of understanding or recognition at the CMGP management level that the key responsibility for a federal program in marine geology is the consideration of the broad range of issues that arise as a result of striking differences in the coastal and marine geologic processes and societal needs in the different coastal areas of the United States.
- The present promotion and reward system for research grade scientists. Scientists are rewarded to continue to produce traditional papers for professional journals. There is no incentive to focus on products more relevant to the mission goals of other agencies.
- There seems to be a great deal of hesitancy among many researchers in CMGP to re-train or embark on new areas of research.

2) Products and Services

A) What are the most valuable products and services that the CMGP presently provides?

- Seafloor maps, transport model results, scientific input on complex coastal issues to resource managers, geologic time perspective on these issues, integrated regional datasets covering areas shared by multiple states.
- Our unique equipment and highly experienced scientific and technical staff allow solving coastal problems in a timely manner that would not be available to other labs or universities. Our participation in the Navarre Island reconnaissance is an outstanding example. Our mapping products show present orientation of the coastal system, and scientific papers help describe the processes. One of the most exciting things about being in the coastal field is that the great unifying discoveries are yet to be made; and so as a research area there is lots of room for growth.
- Digital maps that, in addition to geological research, can be used as base maps by other scientific disciplines and for management decisions. Digital databases.
- Probably its maps and images of underwater topography. However, there appears to be a lack of interest in the interpretation of this kind of information. There is no such thing as an underwater geomorphologist here.
- The most valuable products and services may not be those that are currently provided. The CMGP has well-received efforts in public outreach (e.g., the Ask-a-Geologist activity) and has developed a wide range of activities to meet the needs of some coastal states and other federal government agencies on a restricted range of projects. Because of the loss of core competency, however, the CMGP is less able to

provide the marine perspective to support other programs in the Geologic Division.

- Site-specific studies that assist other federal and state agencies: Palos Verdes, Grand Canyon, etc.
- Building, maintenance, and assessment (i.e., use) of long-term databases on a regional and national scale which is critical to maintaining quality.
- Field capabilities that can be used cooperatively by others (sidescan, oceanography, sampling, high-resolution seismics, data processing).
- I like to think our most valuable product and service is to answer scientific questions that are important basic contributions and that have some application to environmental management issues. This work needs to continue to be authoritative and unbiased, consistent with the USGS reputation.
- CMGP has historically been able to provide research in areas that are not covered by industry or academia. Understanding of plate tectonics and continental orogenic processes came from surveys originally designed for petroleum resource studies. Clearer understanding of the offshore processes help define geologic hazards for coastal communities.

B) What new products and services do you believe could and should be developed?

- Community-based sediment transport model like MODFLOW, a coordinated seafloor and subsurface marine imaging center with equipment and staff available to other agencies and academia (a mobile seafloor "observatory"), a web-based national seafloor data clearinghouse, nearshore long-coring capabilities (joint with ODP), an international seafloor management training center.
- I think the web and Internet-based products should be emphasized and given more weight. The web is an immediate distribution medium that is accessible to a wide range of customers not normally aware of USGS scientific products. The public awareness can be of use in providing critical information for policy decisions.
- Maps and reports that are formatted and designed to meet the needs of end-users and stakeholders rather than the traditional science journal articles that only a handful of scientists ever read.
- A much better distribution of our data in many formats, over the web, interactive databases, etc. Develop a site that tells outside agencies and the public just what data we have.
- Basic research should be done in improving or developing new ways to do things such as seafloor imaging, sedimentology, and geochemistry.
- There needs to be improved support for producing basic scientific journal articles and top-level USGS publications in addition to the currently strong activity in outreach products (e.g., web sites and fact sheets).

The products need to reflect more analysis with a little less emphasis on pretty pictures.

- Establishment and maintenance of data layers (GIS format) for coastal and marine knowledge bank on a regional and national scale.
- With the departure of key scientists that have historically advanced marine geologic research, new areas of fundamental research need to be identified as to areas for which CMGP can provide leadership (sediment transport processes, shoreline and delta stability, tsunamis, etc.).
- With the loss of research vessels, CMGP is not keeping up with the current technology in marine surveying. Organizations like MBARI have mapped the seafloor of Monterey Bay with new high-resolution digital systems that do a much better job collecting data than the best data we can provide. We need forward-looking leadership that will allow some moneys to be spent on necessary upgrades for systems we have been using for the last 10-20 years. Further research should be allowed using existing datasets.
- National assessments of coastal and marine geology issues (erosion, offshore earthquakes, tsunamis, minerals, landslides, benthic habitats).
  - — Coastal and offshore geologic map series (shelf multibeam maps).
  - — Research community access to our archive of samples.

C) What impediments do you see to CMGP developing these new products and services?

- Money, vision, aging scientific staff, post-reduction in force, individual preservationism after the reduction in force, scarcity of charismatic leadership.
- The coastal community is a fairly small network of scientists, and we have a tendency to be inbred and focus on issues of concern among ourselves. CMGP needs to make a concerted effort to broaden its customer base. The coral reefs issue is a good example of an issue that has high scientific merit and strong popular support.
- Emphasis is on solving someone else's problems, rather than on conceptualizing and solving new problems. The thrust should be on trying to anticipate the future, rather than on solving problems for which the private sector is likely to be better equipped.
- A lack of a clear message from management that basic science must be the backbone for any of the products produced by CMGP if program credibility is to grow. This lack of support for a balanced program, especially when combined with increasing restriction of the areas of research that are favored by program management, appears to have discouraged staff from returning to the level of productivity that characterized the marine program prior to the reduction in force in 1995.

- Scientific awareness of its managers about what's coming over the horizon in needed knowledge and expertise concerning the whole of the EEZ. Industry is increasingly headed into deeper and deeper water exploring for and intending to produce energy, mineral, and potentially thermal resources. These searches will eventually involve the EEZ, as they have in the past for the outer continental shelf. But, presently, the programmatic focus of CMGP is inward from about the 100-m isobath toward the coast and upward to the surface of the seafloor and its flanking beaches. But this is not where the industrial world is heading with respect to resources and where the urban communities of at least the Pacific Rim are looking with respect to the rupture areas and consequences of offshore earthquakes and tsunamis. CMGP has designed its program thrust to as much as possible bypass confronting these difficult matters, leaving them for other USGS teams to tend to. With respect to this general stance, it is also the policy of CMGP managers to disengage the survey from cost-sharing cooperation with the ocean drilling program—the largest and most scientifically productive earth science program known to me that is supported fully by the NSF and other national members of the JOIDES community.

D) Do you believe that the end users of the products and services should be included in defining the design objectives of new products and services? If so, what mechanism would you suggest for obtaining their input?

- Yes, pay someone to go to them and solicit their input—at least one CMGP end-user liaison for each center.
- Yes. Feedback forms on publications and products are a minimum, but I would support more proactive means of feedback. A public opinion poll showed the public distrusts the term "mitigation," but we continue to use it. The internet is a very logical place to acquire feedback. Town meetings in other programs on highly visible issues have also proven highly effective.
- Obviously, if our goal is communication, rather than the mere production of a "product," we must listen to the understanding that our end users receive from our products so that we can learn to express the big picture effectively. It is essential, however, that we remain focused on our mission and not get distracted by congressmen or their constituents who wave money at us and tempt us to fritter away our time on projects that could just as well be done by academic or private-enterprise scientists.
- Yes, very much so. Involve them from the very start, solicit their input as you keep them informed of your progress, and verify their needs before providing products. USGS scientists should determine how best to conduct the research and interpret the data, but stakeholders should be in-

volved in that we should know how best our science and research can help them to make more informed decisions and set policy and regulations based on sound and credible science.

- Yes. Have the individual responsible for the end products meet with the concerned end users. Make promotion and good evaluations of products depend on high-quality end-user products.
- No. I don't see a practical way to involve the users in all cases; however, in a case-by-case situation, users can and should be involved.
- Yes. Program management should spend more time identifying the interest and need for marine geologic research that would be of direct benefit to other Geologic Division programs and other federal agencies.
- Developing products and services is an iterative process. End users need to be included, but in a balanced way. One effective way to identify and define useful products is by carrying out (paying) cooperatives with the end users in pilot areas and defining products with these users. Cooperatives with NOAA in Stellwagen Bank Mapping, with Massachusetts Water Resources Authority (MWRA) in Boston Harbor, Massachusetts Bay, and with USACE on New York and southern Long Island are recent examples.
- Some of the end users of our data are likely to be contributors of similar data. We have already experimented with the mechanism of co-authoring data and synthesis from different agencies in the same electronic-based publication.
- There isn't any black box solution to "how to," except to include users in every proposal and for managers to reinforce the requirement for contact. But when scientists make the contacts, managers must be responsive and not reject news of needs that aren't "scientifically interesting."

3) CMGP's Niche

A) Apart from the unique technology and instrumentation in the CMGP, what do you see as the special niche that CMGP fills in the USGS and other federal agencies?

- CMGP controls coastal seafloor data that are not available anywhere else. CMGP interpretation of these data gives genetic and dynamic meaning to NOAA bathymetry and shorelines, isolated EPA sediment quality data, USACE, dam/bridge/beach/port/canal impacts, NWS storm data, USGS Water Resources river discharge data, and BRD coastal species survey data. CMGP data and people link separate programs in other agencies such as ODP and chemical oceanography at NSF. Because most CMGP scientists come from backgrounds outside traditional marine geology, they interface exceptionally well with individuals from a variety of other traditional disciplines (engineers, phytoplankton biologists, and chemists).
- An earth science agency having a strong marine and coastal program is a vision that previous USGS managers should take credit for. It bucks

the trend of one agency/one area. Clearly cooperating with NOAA and the Navy are crucial to our survival as a coastal institution, but the recent cooperation with NASA as a means to prove a new technology shows the unique role CMGP can play. CMGP must be ready to take advantage of these highly competitive opportunities, which means it must accept some failure in trying new concepts.

- Because we focus on a region and because it is expensive to return to a marine region repeatedly, we are more in the habit of interdisciplinary work than are programs that focus on particular hazards or processes on land. This gives us an advantage in scientific integration and synthesis.
- In the federal government the USGS is the only agency with the scientific resources to understand earth science information in context and thus to produce scientifically sound interpolations, extrapolations, and scenarios based on contingencies. The government as a whole should be able to turn to us for the scientific understanding that is needed for their work, and they would, if we proved ourselves capable of communicating effectively.
- The processes in the oceans, both coastal and blue-water, are diverse, complex, and highly interrelated. The broad base of scientific and technical expertise available in CMGP allows us to create diverse teams that can investigate the coastal environment as a total process, enabling us to model interactions among the geologic, chemical, and fluid processes. Single-discipline studies are too limited and are even losing popularity in the outside research community.
- CMGP should be the lead agency of the federal government in dealing with the interactions of the hydrosphere and geosphere. It is unfortunate that other agencies with more money have taken over so many of these marine responsibilities. The USGS needs to assert itself.
- Most of the Geologic Division programs (in geologic hazards, mineral resources, surficial processes, global change, and others) can benefit from a strong diversified marine program. None of these other programs, however, can afford to maintain an independent marine operation. Specific niches include, but are not limited to, mineralization at hydrothermal vents, identification of active faults in the nearshore and coastal zone, monitoring of earthquakes and crustal strain in the coastal ocean, and paleoclimate studies of sedimentation in continental slope basins.
- Short-term, focused, and applied projects are important in terms of our overall mission within the Geologic Division. However, I think our identity and niche is defined by our scientific expertise in key fields of marine research. Presently, this is in a state of flux and therefore is hard to define. CMGP currently seems to provide leadership in the areas of sediment transport and geotechnical engineering.
- Long-term: USGS can bring expertise to bear on issues over the long

term (and see monitoring below). Knowledge of regional geology and issues is maintained over the long-term, a 'corporate memory' that is often needed to address issues in the coastal ocean.

— Regional: USGS can apply a systemwide and regional perspective (geologic framework) to coastal issues. Regulatory agencies are often forced to limit the spatial extent of investigations.
— Multidisciplinary: the CMGP brings a multidisciplinary team (geologists, geophysicists, oceanographers, chemists) to address issues.
— Unbiased (i.e., non-regulatory) high-quality science: USGS's non-regulatory function provides the public and other agencies with an unbiased assessment of coastal processes and their effects. This requires scientific excellence.
— Monitoring: USGS is uniquely suited to carry out well-planned long-term monitoring of environmental change in the coastal ocean. USGS regional studies provide the spatial context for these temporal observations, which can improve fundamental understanding of coastal processes. NSF and ONR do not fund these studies.
— Model development and maintenance:

- Expertise in designing, conducting, and interpreting surveys of coastal and marine environments.
  Also a "corporate memory" and a long-term commitment to archive data and to make it available to the public.
- Our niche is large, sustained, regional, and national studies and assessments of coastal and marine geologic issues of concern to national (federal) and regional entities (e.g., state, industry, or private coalitions). CMPG has the unique skills and tools used in the study of geology in subaqueous environments, skills and tools that transgress multiple aspects of geology in the marine environment.
- By its entrepreneurial and dynamic nature, CMGP has the potential ability to network with many organizations to build coalitions and thematic programs.

B) Do you believe that CMGP should and could work more collaboratively with other USGS programs and divisions?

- Yes, although CMGP currently does better at this than most other USGS programs and divisions.
- Yes, and in cooperation with state agencies, too.
- Yes, most definitely. Despite the rhetoric to the contrary, there are still a great many difficulties in doing this.
- We should definitely increase our links with Water Resources and the earthquake, volcano, and mineral communities.

- Yes, very much so, especially in terms of marine mineral studies, seismology, geophysics, and habitat studies.
- Collaboration is an ideal to strive for, but is difficult in practice, especially in the USGS, where turf still matters greatly.
- There is tremendous pressure to work collaboratively inside and outside USGS. Many coastal science issues benefit from broad collaboration, often increasing the skills brought to bear on a problem and effectiveness of the project. However, cooperation for cooperation's sake is non-productive, and building and nurturing cooperative partnerships can take a tremendous amount of effort. Cooperatives are often most successful if built at the PI level. Management can facilitate these cooperations by identifying opportunities.
- Yes. An obvious partner is BRD, but they have few marine biologists, to my knowledge. There is probably some opportunity to collaborate in studies of estuarine environments. The most fruitful collaboration might occur in the Great Lakes region, where BRD has a large commitment and where CMGP can apply its expertise in the same way it does in coastal environments. Expansion of activities to the Great Lakes region would require additional funding and personnel so that ongoing programs are not weakened.
- CMGP should be integrated into the major programmatic focuses on resources, hazards, environment, and the growth of knowledge. As noted, coastal and marine studies do not in themselves constitute a program effort.

C) What specific suggestions do you have for best filling CMGP's specialized niche or making use of CMGP's unique abilities?

- More staff in Reston and the Washington area, separate branch manager roles from inter-agency liaison (salesperson) roles; identify areas of strongest overlap with other agencies and transfer people duplicating USGS efforts there to USGS (or vice versa); increase size of each branch, or else localize one major specialty in each branch (e.g., seafloor mapping in Woods Hole, geochemistry in St. Pete, something else in Menlo).
- I think we should make it easy for the science public and the general public to know our capabilities and successes. The public access literature we have has been generally good but needs to be strengthened. More important our presence at national meetings at groups other than coastal scientists is needed. Town meetings on specific topical coastal issues were mentioned before. Our web presence should be encouraged to grow.
- Encourage the diverse nature of our scientific and technical bases. Do

not allow us to become an agency with only a few disciplines or scientific programs. We should even keep some expertise in blue-water processes.

- Get back to basic research as the main theme. Perhaps design a two-tiered system in which part of the staff is assigned to basic research and another part assigned to customer problem solving. Recognize that at least five years is required to get research results. Review assignments on a five-year interval. Reassign staff based on productivity and attempt to satisfy employee needs and desires.
- The Geologic Division needs to complete implementation of the management structure that was set up in 1996 after the RIF. Under the revised division structure, program coordinators are supposed to be seeking new areas of application for the unique capabilities of their programs and to identify new collaborative efforts (refer to the position descriptions that are currently approved). Unfortunately, some program coordinators are spending most of their time on managing expenditures even though this function is not part of the job description (expenditure management resides with the identified line managers [i.e., the chief geologist, regional geologists, and team chief scientists]). If the Geologic Division were to fully implement the new management structure, cross-program collaboration would be facilitated because overall program direction would more closely reflect customers and potential collaborators unique to each region.
- It is important that CMGP clearly define its role with respect to marine mineral and energy resource. This role should be defined in the context of a working relationship with other government agencies, academia, and industry. All of these are potential partners, collaborators, and clients, and they should be part of the planning process for future program activities.
- Field Centers: The CMGP has co-located field centers adjacent to major centers of geologic and oceanographic expertise to share specialized facilities, facilitate cooperation, and to create a stimulating intellectual climate. These centers have unique operational needs not experienced by programs that are located in national centers (network support, libraries, administrative support, etc.) that often need special attention. The USGS should consider co-locating other activities at these existing centers that would strengthen cooperation or take advantage of unique capabilities as opportunities present themselves.

4) CMGP's Organization

a) Weigh on a scale of 1 to 10 the effect of the following on the CMGP's ability to successfully address the problems, products, and services you have listed in questions 1 and 2. Please explain your reasoning.

b) Do you have any suggestions on alternative methods of managing these issues listed below?

Resource (people and products) sharing between field centers, GD programs, and other USGS divisions

- Generally good but not consistent in all projects.

  Idea: organize something like departments in CMGP so that sediment transport folks in Menlo, for example, have some regular contact with sediment transport folks in Woods Hole, and geochemists can work on coordinated analytical arrangements (national contracts or shared equipment acquisition rather than every branch for itself).
- I don't think sharing among field centers, GD programs, and other USGS divisions is the issue, but I think communications between these parts of the programs is crucial. During Hurricane Mitch, mapping had an effort, CINDI, to respond by producing historical map products and sending a field team. I could not participate in these discussions because of my location in a field center, yet we were extremely close to the action and had lots of experience in this arena. Impediments to communication continue to hurt us in all areas.
- The big picture is a massively cooperative effort. Leadership should rally the USGS around its common mission, instead of throwing dog biscuits to projects that cross-organizational boundaries.
- The CMGP staff is rapidly aging, very few people are brought in, and since the RIF most technical help has disappeared. When I run cruises or do fieldwork I am forced to ask high-grade staff to act-in support roles because there are so few support staff available. This is very non-cost efficient, both for me as a project scientist and for the USGS. We need to be able to share resources without having to "pay a ransom" to do so. Our multidisciplinary research would appear to be the ideal medium by which we should be able to share resources; however, politics and egos constantly get in the way of this.
- Designate specific centers (Menlo, Woods Hole, or St. Pete) as the program office that has the expertise in specific technical or research skills. When a CMGP project is staffed, draw the required expertise from that center (instead of from the location of the project). We are doing this in a small way with our extensive collection of physical oceanographic equipment. Much of this equipment is readied for the field in the Woods Hole office, then sent to programs at all centers.
- Other GD programs that asked for data from CMGP were told to provide their own employees to work on the data and that CMGP would not

provide assistance. This response has affected several projects in both CMGP and other GD programs.

Cost of CMGP products and services to outside clients

- They are getting a bargain! In general, the cost to them should actually be much higher.

  Idea: identify more end users before development of products and get more contributions up front.

- Given that most of Dr. Gilberts' time is booked up I think our charges are about right, or could be higher. I would be reluctant to get involved with the high costs of large oceangoing research vessels.
- Our working capital fund should be expanded to handle outside journal publications.

  If we could sell our digital products and place the money into a general working capital fund for future publication costs, more money would be freed up for operating expenses.

- The federal government needs the big picture to inform its actions. It should be provided to clients outside the federal government for the simple costs of media and reproduction.
- Cooperatives—shared costs should be emphasized.
- If CMGP is to provide products and services to outside clients, the costs must be competitive with the private sector. Because there are so many specialists in the private sector, it may be difficult for CMGP to beat their prices. Instead of focusing on sales, CMGP should work on new ideas and the implementation of these ideas.
- Our costs to outside clients can be low if we can set up a cost-share program with them. If an outside client was required to post the true and entire cost of a USGS project, I doubt we could compete with an outside contractor. However, the product might be superior.
- Most data distribution is through NGDC, which charges relatively high costs for reproduction. Researchers that contact CMGP directly often are given the data with no charge, as no method of billing is available. Offshore mapping performed with our own vessels was a bargain compared to the current ship-leasing methods.

Hiring procedure and criteria (e.g., world-class, specialty, term appointments)

- System is byzantine. Most hiring appears to be fairly convoluted and is done as a stopgap to meet commitments already in place. Recruitment of experienced scientists to expand into a new area or give leadership to a growing group seems rare. Many good people are brought in as terms and contractors but leave before their contracts or terms expire because there is no future with the organization. This is a very inefficient use of training time and money.

- The current pattern of hiring research scientists as permanent employees, while support staff are temporary, is completely backwards. Research staff are interchangeable and come ready-trained from the university. Support staff, especially technical types, perform unique jobs, require extensive on-the-job training and, on leaving, take critical skills and knowledge with them.
- The research staff needs to be better balanced to accomplish our missions. Specialists that retire or leave are being replaced by others in different disciplines. The strategy for replacement of specialists seems to be very shortsighted and will probably hurt us in the long run. We desperately need more technical support to handle the lower-grade aspects of research, yet the problem only seems to get worse.
- Except for world-class scientists, hire most research staff on a term basis. Convert the ones that work out (and are needed for long-term project goals) to permanent. Hire technical staff needed for long-term projects as permanent. Replace staff that retires, but perhaps not in exactly the same field.
- In the end, term appointments will lead to disaster. Eventually there will be no corporate memory and no legacy and no loyalty.

Staff personal growth opportunities

- Many and varied, especially because of the association of field centers with university and other nearby research institutions.
- Getting there from here will require us all to learn new skills and attitudes. Training should be provided for tasks that employees would need to perform in the future, not simply for tasks they are performing already.
- Growth opportunities seem minimal. The older staff is topped out and there has been only a minimum of new hires.
- This is a major problem. Ideas from the staff are not appreciated unless they fit into management's view of what the problems are.
- Mostly self-driven, but available. In-service training requirement might prepare research staff for career adjustments needed with changes in research directions.

Staff evaluation criteria and reward system

- Compensation is competitive, although advancement seems to be tied to a fairly rigid federal grade system that restricts ability to reward shining stars or remove deadwood.
- Currently, research scientists are rewarded for emulating academic scientists. Support staff are rewarded for supporting the projects of the research staff. Nobody is rewarded for furthering the CMGP's unique mission. Obviously, this must change.
- The survey still does not have a reward system that encourages scientists to support USGS goals. The research staff is mainly rewarded for individual products (i.e., scientific papers).
- This issue has been addressed in the last two years, but there has been little evidence that any of the recommendations have been implemented.
- Current evaluations tend to be meaningless, as there is currently just a pass-fail rating. All employees can work at the pass level, leaving little incentive to outperform. Rewards are virtually nonexistent except for a select few. Positions are described as non-promotable, removing incentive by employees to excel.

Staff recognition outside USGS

- Good overall but variable. Opportunities exist to advance professional reputation, but not all scientists take advantage of them.
- Encourage journal articles, talks at other agencies, and high-quality scientific work. Publicize our large projects in *EOS* or in other scientific newspapers.
- Not considered important to CMGP management, although it should be, unless the recognition is from a full-paying client.
- Given the way things are going toward customer-driven research, there will be a diminishing of recognition outside the USGS. Workers will be recognized by customers, but that recognition may be short lived if the prices are too high.
- In the fields with which I am most familiar, the staff reputation was unquestioned; with the losses during the reduction in force and with the continuing attrition since then, however, it is becoming a problem.
- Recognition of CMGP staff by outside agencies and academia has noticeably declined since the reduction in force.
- Quality and quantity of CMGP data have suffered in the last four or five years. Perhaps this current review will allow for an improved work atmosphere, an increase in productivity, and a resulting improvement in professional recognition.

Publication policies

- These policies seem to be in a state of flux from a traditional USGS system to a system open for interpretation by each scientist and program. The traditional internal publication system seems to be dying; web-based products and fact sheets are thriving, but support for external products is not always available.
- I think this is an oxymoron. The two concepts do not mutually coexist here. Having a resource like the computer center here with our plotters is a tremendous resource and has done much to improve productivity and original thinking. Much credit goes to Rob Wertz. Most of the other 'publication policies' have been hurdles to overcome.
- Pay all page charges for outside journals. Either pay outside agencies to publish our maps, or get our group to do this better. Publish glossy paper products (like atlases) with just a few pages and put the extensive products on a CD.
- Each project should produce scientific papers and outreach products. The USGS motto, "Science for a changing world," has come to mean consultants; the science has been abandoned for what is viewed as immediate societal need that can be demonstrated to a senator or congressperson.
- Publication policies have been addressed in a recent memorandum. The division seems to be on a reasonable track. My personal bias is that outside publication should be encouraged so that any basic research that is done can compete with any new and innovative research that comes from other sectors.
- A lot of attention has been put on creating posters, web pages, etc., with attention-grabbing graphics; less attention to scientific content. The scientific content has been poor on many posters seen this year at AGU. A publication with data that are unprocessed and misinterpreted undermines the professional reputation of the organization.

APPENDIX D

The Relation Between the USGS Geologic Division Goals and the Coastal and Marine Geology Program

REVIEW OF THE GEOLOGIC DIVISION'S GOALS AND THEIR IMPACT ON CMGP

The recently published document entitled "Geology for a Changing World" (USGS, 1998h) presents a science strategy for the years 2000-2010 for the Geologic Division of the USGS. The document identifies seven science goals aimed at addressing pressing issues facing the nation in the next decade, and they include traditional areas of research for the Geologic Division, as well as societal issues that are becoming increasingly important due to concerns about the preservation of the environment and the quality of life.

USGS Geologic Division Strategic Science Goals, 2000-2010

1. Conduct geologic hazard assessments for mitigation planning.
2. Provide short-term prediction of geologic disasters and rapidly characterize their effects.
3. Advance the understanding of the nation's mineral and energy resources in a global geologic, economic, and environmental context.
4. Anticipate the environmental impacts of climate variability.
5. Establish the geologic framework for ecosystem structure and function.
6. Interpret the links between human health and geologic processes.
7. Determine the geologic controls on groundwater resources and hazardous waste isolation.

Within the Geologic Division, a number of programs exist to conduct the studies necessary to accomplish these goals. Goals 1 and 2 clearly justify the need for hazards programs (earthquake hazards, volcano hazards, landslide hazards, and the global seismic network). Minerals and energy resources programs address Goal 3, and the earth systems dynamics program is aimed at Goal 4. The geologic mapping and ecosystems programs address aspects of Goals 5 and 6, and mapping is an important component of studies to address Goal 7, although there is no explicit program to address Goal 7.

The National Cooperative Geologic Mapping Program (NCGMP) and the CMGP stand out as not being *explicitly* tied to a GD science goal. The NCGMP clearly provides the basic framework for almost all geological studies, and hence it encompasses mapping at a variety of spatial and temporal scales to address many of the science goals. The CMGP is focused on the marine and Great Lakes geographic regions, hence the studies that are conducted under its auspices span all GD science goals.

GD Scientific Programs

Program	FY99 Budget
Hazards	
• Earthquake	$50.4M
• Volcano	$19.8M
• Landslides	$2.4M
• Global Seismic Network	$3.8M
Resources	
• Minerals	$62.7M
• Energy	$26.0M
Earth Systems	
• Geologic Mapping	$22.5M
• Coastal and Marine	$38.2M
• Ecosystems	$2.6M
• Earth Surface Dynamics	$13.6M

RELATION OF CMGP THEMES AND SUBTHEMES TO THE GEOLOGIC DIVISION SCIENCE GOALS

In 1994, the USGS implemented a five-year CMGP that outlined proposed studies to understand the coastal and offshore areas of the United States. This plan was modified in 1997 to take advantage of new opportunities and issues and to account for changes in budgets and staffing.

The stated mission of the CMGP of the USGS is to "provide the nation with objective and credible marine geologic science information based on research, long-term monitoring, and assessments." The program is designed to (i) describe marine and coastal geologic systems, (ii) understand the fundamental processes that create, modify, and maintain them, (iii) develop predictive models that provide an understanding of natural systems and the effects of human activities on them, and (iv) provide a capability to predict future change. The 1997 Five Year Plan identifies four themes, each with 2-4 subthemes, as the focus of investigations by CMGP.

Coastal and Marine Geology Program Themes and Subthemes

Theme 1: Environmental Quality and Preservation

To understand sediment and pollutant erosion, transport and deposition, fragile environments, the importance to the nation of sea- and lake-floor environments as biological habitats, and as record keepers of long-term environmental change.

Subtheme 1: Pollution and Waste Disposal
Subtheme 2: Fragile Environments
Subtheme 3. Marine Reserves and Habitats

Theme 2: Natural Hazards and Public Safety

To better understand the frequency and distribution of catastrophic events that elicit federal response (storms, earthquakes, and landslides), the geologic processes acting in the affected marine and coastal regions (e.g., coastal erosion), and the local and regional susceptibility to change.

Subtheme 1: Coastal and Nearshore Erosion
Subtheme 2: Earthquakes, Tsunamis, and Landslides

Theme 3: Natural Resources

To develop and extend understanding of the formation, location, and geologic setting of offshore mineral and petroleum resources, the geologic effects of resource extraction, and how offshore resource occurrence can help in the search for analogous onshore deposits of economic significance (could address areas outside the EEZ).

Subtheme 1: Water Resources (Coastal Aquifers)
Subtheme 2: Marine Mineral Resources
Subtheme 3: Energy Resources

Theme 4: Information and Technology

To develop and maintain a national comprehensive source of multidisciplinary data and information that can be easily accessed and used by government policymakers, research scientists, and the public, and to maintain scientific instrumentation and platforms necessary to carry out research and mapping activities.

Subtheme 1: Systematic Mapping of the Coast and Seafloor
Subtheme 2: Coastal and Marine Information Bank
Subtheme 3: Assessments and Evaluation of the Information Bank
Subtheme 4: Technology and Facilities

The Committee to Review the USGS Coastal and Marine Geology Program compared the CMGP themes and their stated goals with the GD's seven strategic science goals. The GD goals approach the geological studies of the environment from the "impact or change" (whether natural or anthropogenic) perspective, which results in a disconnection between them and the areas of interest in CMGP Theme 1. Environmental Quality and Preservation (Theme 1) consists of a collection of unrelated topics. Studies of long-term environmental change through the sedimentary record do not truly address issues of environmental quality and preservation. Sediment erosion, transport, and deposition are naturally occurring processes rather than environmental quality and preservation issues, unless disrupted by human activity. The transport and fate of pollutants is an environmental quality issue and supports GD Goal 7. The preservation of fragile environments and biological habitats clearly depends on the successful conduct of GD Goal 5, but their health can also be affected naturally by climate variability (Goal 4) or by human activity. The role of the CMGP in conservation of marine and coastal areas as marine sanctuaries is one of seafloor mapping of bottom morphology and biological habitats for management purposes, which fits better under Theme 4, Subtheme 1: Systematic Mapping of the Coast and Seafloor. The committee is concerned that activities in Theme 1 do not provide a coherent scientific program. If it is recognized that GD Goals 4 and 5 require an understanding of the dynamic geologic and biologic systems that interact in the coastal and marine environments, then a revision of Theme 1 to reflect the need for systems studies would encompass the range of environments and processes already included in the stated objective.

Natural Hazards and Public Safety (Theme 2) clearly addresses GD Goals 1 and 2 and has consequences for Goal 6, so there appears to be an appropriate match. The GD approaches geological studies of the environment from the "hazard, impact or change" (whether natural or anthropogenic) perspective, which fits well with CMGP Theme 2. However, in the GD, there are four other programs that address hazards of various types (the earthquake, volcano, landslides, and global seismic network programs), and how the CMGP efforts dovetail into the larger efforts by these other programs is a concern of the committee.

Natural Resources (Theme 3) encompasses and directly reflects GD Goal 3. The committee noted that the statement of the scope of studies in Theme 3 omits investigations related to Subtheme 1: Water Resources (Coastal Aquifers), which add another dimension to the theme. However, the relative roles that the CMGP and the USGS Water Resources Division should play in studies of water resources and intrusion of saltwater into coastal aquifers is unclear. If the main effort of the CMGP is to provide the geologic framework of the coastal region so that the Water Resources Division can assess the impact of saltwater intrusion on freshwater resources, that effort could be better included in Theme 4, Subtheme 1: Systematic Mapping of the Coast and Seafloor. If the CMGP intends to include detailed studies of subsurface fluid flow in the nearshore area, then investigation of water resources needs to be included as a goal of Theme 3. Another important issue is how CMGP studies integrate with those of the Minerals and Energy Resources Program in the GD.

Information and Technology (Theme 4) and the four subthemes in it address a series of infrastructure issues rather than a scientific or environmental theme. It focuses on the responsibility of the CMGP to manage the information and data that are collected and to disseminate them in a form of use to the public and to policy- and decisionmakers. This theme addresses the operational objectives of the GD, which were put in place to improve the usefulness and accessibility of information and to promote the flexibility and vitality of the staff. The six operational objectives are:

- Greatly enhance the public's ability to locate, access, and use GD maps and data;
- Maintain a first-rate earth-system library;
- Effectively transfer the knowledge acquired through GD science activities;
- Promote vitality and flexibility of the scientific staff;
- Promote interdisciplinary research; and
- Institute internal and external reviews.

Clearly, information management and dissemination, maintenance of scientific instrumentation, and access to platforms represent important functions of CMGP that must be maintained as a critical operational component of the program. However, the committee recommends that the themes and subthemes should be limited to scientific and environmental issues in the coastal and offshore regions.

APPENDIX E

Summary Budget Data

Coastal and Marine Geology Program FY99 Program Budget Overview

Category	Costs ($ million)
Salaries	10.2
Project Operations	7.8
USGS Assessment and Set-aside	7.6
Team and Center Infrastructure	6.0
Team and Center Assessment	3.0
Space Costs	2.1
Program Office	1.6
TOTAL	38.3

U.S. Geological Survey Coastal and Marine Geology Program FY99 Funding to Science Projects by Theme

Theme	Subtheme	Costs ($ million)	Theme Total ($ million)
Theme 1—Environmental Quality and Preservation	Pollution and Waste Disposal	3.1	
	Fragile Environments	1.2	
	Marine Reserves and Habitats	1.8	
	Fundamental Environment	1.0	7.1
Theme 2—Natural Hazards and Public Safety	Coastal and Nearshore Erosion	3.3	
	Earthquakes, Tsunamis, and Landslides	2.4	
	Fundamental Studies	0.7	6.4
Theme 3—Natural Resources	Water Resources	0.2	
	Marine Mineral Resources	0.7	
	Energy Resources	1.1	
	Fundamental Resources	0.03	2.03
Theme 4—Information and Technology	Mapping of the Coast and Seafloor	0.9	
	Information Bank	0.6	
	Assessments	0.2	
	Technology and Facilities	0.7	2.4
TOTAL			17.9

Bureau Context: Funding USGS Appropriation FY99

Bureau	Funding ($ million)	Percent
Biological Resources	162.5	21
National Mapping	138.3	17
General Administration	27.3	3
Facilities	21.5	3
Geologic	201.2	25
Coastal and Marine Geology Program	38.2	5
Water Resources	209.0	26
TOTAL	798.0	100

Comparison of CMGP FY99 Funding to Other GD Programs

GD Programs	Funding ($ million)
Landslide Hazards	2.4
Ecosystems	2.6
Global Seismic Network	3.8
Earth Surface Dynamics	13.6
Volcano Hazards	19.8
Geologic Mapping	22.5
Energy Resources	26.0
Coastal and Marine Geology	38.2
Earthquake Hazards	50.4
Minerals Resources	62.7
TOTAL	242.0

Appendix F

The U.S. Geological Survey Coastal and Marine Geology Program

Program Planning, Decision Process, and Operations

S. Jeffress Williams

PROGRAM PLANNING—FIVE-YEAR NATIONAL COASTAL AND MARINE GEOLOGY PLAN

A National Coastal and Marine Geology Program Plan was initially prepared and implemented in 1994 and updated and revised in 1997, following staff and budget adjustments. The current plan serves as a combination strategic and operational plan used to manage and guide research in the Coastal and Marine Geology Program (CMGP). The 1994 Plan framework was the product of numerous meetings in 1993 involving most U.S. Geological Survey (USGS) researchers involved in coastal and marine geology research with additional input from external partners and clients (e.g., coastal state and territory geological surveys, other federal ocean agencies [NOAA, NSF, NASA, ONR, MMS, EPA], academia, NGOs, and industry).

The CMGP focused on four primary themes (Environment, Natural Hazards, Resources, Information and Technology), and the final 1994 version was prepared by a team consisting of the managers from the three centers, senior science staff from the three centers, and the CMGP. The Plan provided structure for projects and outlined specific future project directions and science priorities. Project budgets were indicated at two levels, full and partial implementation, depending on the final budget appropriations from Congress.

The 1997 revised Plan, prepared by the CMGP to reflect new realities of smaller staff, level-to-declining budgets, and new research directions in marine geology, was based on input and suggestions from the USGS research staff and written materials from other federal agencies expressing their interests, emphasis, and recommendations for where the CMGP should focus in marine geology research. The Plan provided structure for existing and future projects and suggested areas and topics for future research emphasis. The four themes were retained, but specific budget details were not included because of the uncertainty of future appropriations from congress.

Following the NAS/OSB review of the CMG Program in 1999, a new National Program Plan will be prepared to conform to the science goals and operational objectives detailed in the new Geologic Division (GD) Science Strategy, "Geology for a Changing World."

BUDGET PLANNING

The annual budget for CMGP is prepared by the Program office, based on department, bureau, and division science priorities, marine science priorities of Congress, the National Plan, and inputs to the Program office from a wide variety of internal and external sources. Budget emphasis at the theme level has remained as discussed in the current National Plan. Overall budget guidance is given to the Program office by the division, bureau, DOI, and OMB offices. The Program office in 1998 maintained budget balance between the themes discussed in the 1994 Plan and revised in the 1997 National Plan. The "Green Book" document fully describes for Congress the marine science priorities and major accomplishments for CMGP. It contains some information on the distribution of research funds across the four themes and describes implications of a $3.5M funding reduction contained in the President's budget the FY2000.

DEVELOPMENT OF THE ANNUAL PROGRAM PROSPECTUS

The Program Prospectus is developed each year by the Program office to provide guidance to the USGS research staff on the distribution of the funds for salary and operating expenses based on the current National Plan and the expected budget appropriation. The research staff responds to the Prospectus by preparing written "workplans" for continuing projects or "proposals" for new research projects. The Prospectus contains suggested figures for both salary and operating expenses for projects. This level of detail is included so as to ensure a healthy operating margin for CMGP. In many cases lead PIs were identified based on science expertise, leadership skills, and availability. The Prospectus also requested workplans for center assessment activities and activities supported by direct-funded overhead costs at the three centers.

The annual Prospectus is developed based on inputs from many sources:

- Knowledge of the USGS, DOI, OMB, Congress funding priorities;
- Meeting science needs of other DOI bureaus (e.g., NPS, MMS, FWS);
- Meeting the science needs of coastal states/territories;
- Response to division planning and meeting the new science goals;

• Congressional requests (e.g., Los Angeles subsidence, South Carolina and Georgia erosion, coastal hazard risk assessment);

• Direct inputs from other federal agencies (e.g., NOAA, NSF, FEMA, EPA, ONR, and NOPP);

• Interactions with other USGS divisions, such as ground water, marine habitats, and coral reefs;

• Program Advisory Council (PAC) recommendations;

• Interactions with other GD programs (e.g., Ecosystem, Climate, Energy, Earthquake, and Minerals);

• Mid-year project briefings and discussions;

• USGS research staff comments and recommendations;

• Knowledge of staff interests, capability, and availability; and

• Knowledge of the future directions and priorities of marine science (FUMAGES).

The PAC serves as a science review panel made up of CMGP research staff as well as scientists from other inside and outside the USGS. The purpose of the PAC is to provide broad scientific review, recommendations, and advice to the Program office. Their primary task is to review the scientific quality and program relevance of continuing workplans and new proposals submitted each year to the Program office for funding.

EVOLVING COMPOSITION OF THE PAC

For the FY98 review, the PAC had 16 members providing scientific and geographic coverage and they were tasked with evaluation of scientific value, program relevance, and productivity of the projects funded by CMGP:

- 5 research scientists from the Western Team, Menlo Park;
- 3 research scientists from the Eastern Team, Woods Hole;
- 2 research scientists from the Eastern Team, St. Petersburg; and
- 6 managers (2 Western Team, 2 Eastern Team, and 2 CMGP).

For the FY99 review, the PAC had 16 members with the same basic task:

- 4 research scientists (Western Team, Menlo Park);
- 3 research scientists (Eastern Team, Woods Hole);
- 2 research scientists (Eastern Team, St. Petersburg); and
- 1 research scientist (Western Energy Program Team).

The 6 Managers (2 Western Team, 2 Eastern Team, and 2 CMGP) made staffing and funding evaluations.

For FY 2000, the PAC consists of 17 members with the task of providing science review of the proposals and workplans and recommendations to the Program office:

- 4 research scientists (Western Team, Menlo Park);
- 2 research scientists (Woods Hole Team);
- 2 research scientists (St. Petersburg Team);
- 1 research scientist (Western Energy Program Team); and
- 2 external research scientists (Eastern Geologic Mapping Team and Biologic Resources Division).

CMGP ANNUAL PROJECT BRIEFINGS

To keep abreast of project activities, progress, and accomplishments and in anticipation of developing the Program Prospectus for the following year, the PAC organizes a series of project briefings in January/February. These briefings provide the Program office and the PAC members with an understanding of how projects are performing and what concerns and issues need to be considered in developing the Prospectus and in preparing for the next year's workplan and proposal cycle.

The briefings are organized by the PAC and held at each of the three centers. They consist of one to two days of presentations to PAC members, to Program office staff, and to center managers. The presentations are open to staff. The presenters are asked to focus on current research activities and recent accomplishments and products, as well as issues and concerns of the project PI and staff. Written comments, recommendations, and feed back from the project briefings are developed by the Program and the PAC and given to the project PIs.

ANNUAL CALENDAR FOR SCIENCE PLANNING IN THE GEOLOGIC DIVISION, INCLUDING CMGP

Early March—The Prospectus is made available through a division-wide release to the USGS research staff.

Early April—Preproposals for new project starts are submitted by USGS staff.

April—All of the preproposals are reviewed by the division program coordinators, and based on scientific merit the preproposals are rejected or authors are encouraged to submit full proposals.

Early June—Workplans for continuing projects and new project proposals are submitted.

June-July—The Program Advisory Council and Management Team meet together to review the proposals and workplans. The PAC evaluates and comments on scientific merit and Program relevance. Written comments on scientific merit, relevance, and productivity are developed by the PAC during and following the meeting and are then reviewed by the Program office and finally sent to the PI's. Managers meet at this time to discuss staffing and funding issues for both science projects and for direct and assessment-funded Program overhead projects.

August—Initial funding decisions for the upcoming FY are developed by the Program office based on the PAC recommendations, management discussions, and the anticipated budget. Unsuccessful proposals and underfunded staff are identified and discussed.

September-December—The PI's, and Center Managers respond to PAC comments and draft funding decisions. The Teams and Regions provide revised workplans and staffing based on the Council's comments and the Program's preliminary funding decisions. The Program office learns of congressional appropriation decisions and adjusts draft budget, if needed, in consultation with the center managers. CMGP funds are then distributed through the division to the scientists.